Ti = 50	Zr = 90	? = 180
V = 51	Nb = 94	Ta = 182
Cr = 52	Mo = 96	W = 186
Mn = 55	Rh = 104,4	Pt = 197,4
Fe = 56	Ru = 104,4	Ir = 198
Ni = Co = 59	Pl = 106,6	Os = 199
Cu = 63,4	Ag = 108	Hg = 200
Zn = 65,2	Cd = 112	[illegible]
? = 68	Ur = 116	Au = 197?
? = 70	Sn = 118	
As = 75	Sb = 122	Bi = 210?
Se = 79,4	Te = 128?	
Br = 80	J = 127	
Rb = 85,4	Cs = 133	Tl = 204
Sr = 87,6	Ba = 137	Pb = 207
Ce = 92		

THE ELEMENTS

AN ILLUSTRATED HISTORY OF THE PERIODIC TABLE

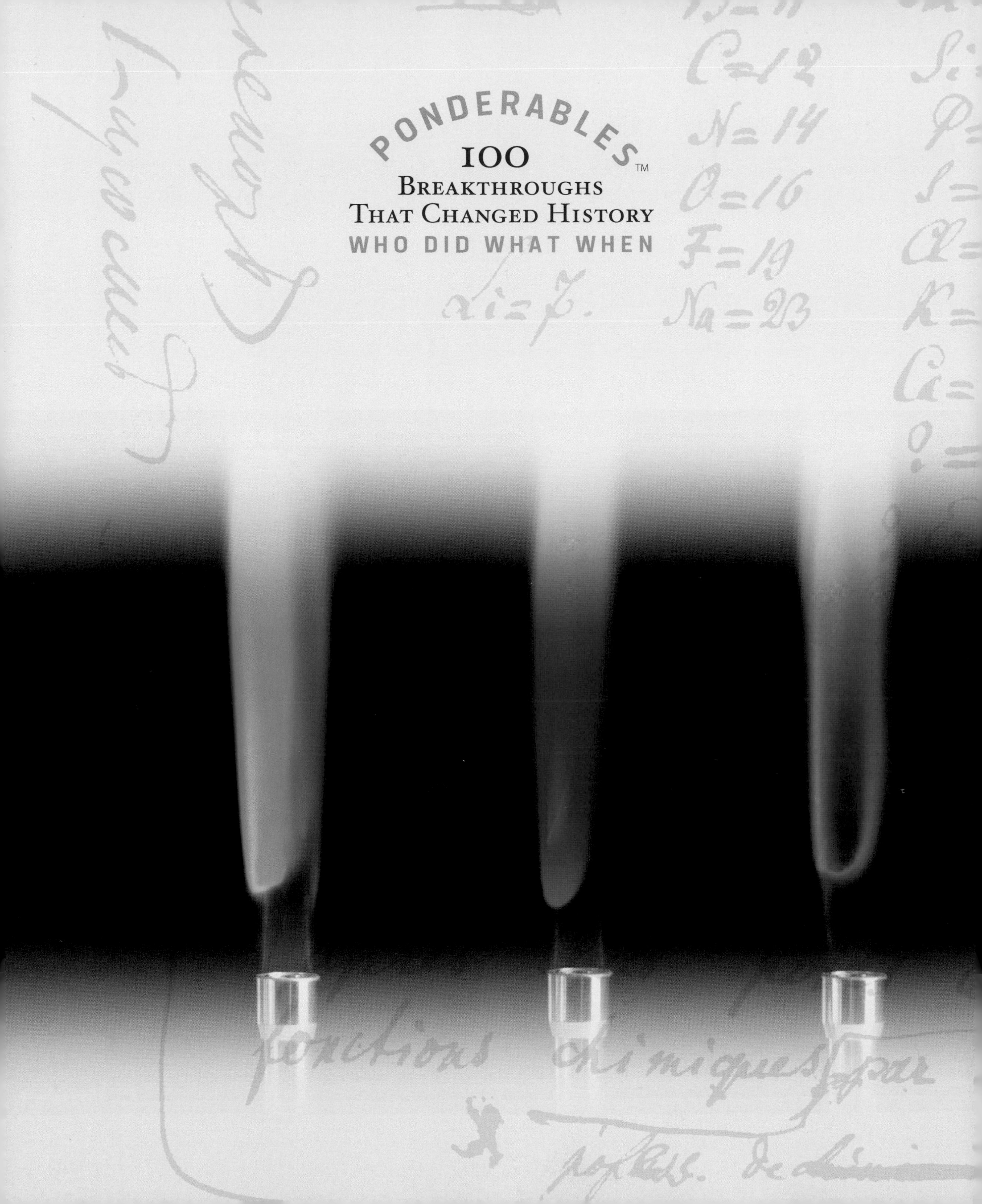
PONDERABLES™
100
Breakthroughs
That Changed History
WHO DID WHAT WHEN

THE ELEMENTS

AN ILLUSTRATED HISTORY OF THE PERIODIC TABLE

Tom Jackson

SHELTER HARBOR PRESS
NEW YORK

Contents

THE SCEPTICAL CHYMIST: OR CHYMICO-PHYSICAL Doubts & Paradoxes, Touching the SPAGYRIST'S PRINCIPLES Commonly call'd HYPOSTATICAL; As they are wont to be Propos'd and Defended by the Generality of ALCHYMISTS.

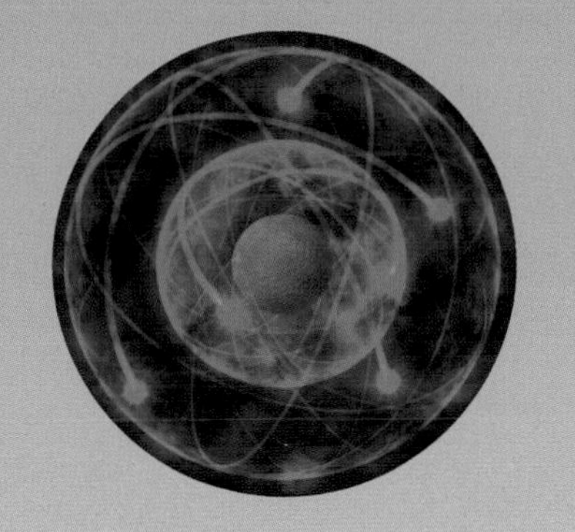

THE 19th CENTURY: THE GREAT AGE OF SCIENCE (1800–1900)

THE MODERN ERA (1900–PRESENT)

Introduction

OURS IS VERY MUCH A MATERIAL WORLD. HAVE YOU EVER WONDERED WHAT IT IS MADE OF, WHAT IS IN IT DEEP DOWN? IF SO YOU ARE NOT ALONE, AND CENTURIES OF PONDERING HAVE COME UP WITH A GREAT MANY ANSWERS OVER THE YEARS.

Four elements were enough for most of human history.

Alchemists began the process of cataloging elements by their properties.

SCHEMA MATERIALIUM PRO LABORATORIO POR

I	MINERÆ	☄	♁	♛					
II	METALLA	☉	☾	♀	♂	♃	♄		
III	MINERALIA	♁	Bismuth	Zinck	Marcasit	Kobolt	Zaffra		
IV	SALIA	⊖	⦶		○	✳	☿	Borax	Chrisocolla
V	DECOMPOSITA								
VI	TERRÆ	⊕	Crocus ♂	Crocus ♀	Vitrum ♁	Vitrum ♄	Minium Lithargirium	Cadmia Tutia	Ochra Schmalto
VII	DESTILLATA		Sp	Sp ⊖	Sp ⦶	Sp V	Sp ☿		Sp ⊡
VIII	OLEA	Ol	Ol	Ol fixum ☿	Ol p deliq ☿	Butyr ♁	Liquor Silicum	Ol Therebent	
IX	LIMI	C.V. ♆	Arena	Creta Rubrica	Terra Sigillata Bolus	Hæmatites Smiris	Talcum	Granati	Asbestus
X	COMPOSITIONES	Fluxus Niger	Fluxus Albus		Coloriza	Decoctio	Tirapelle		

The thoughts and deeds of great thinkers always make great stories, and here we have one hundred all together. Each story relates a ponderable, a weighty problem that became a discovery and changed the way we understand the world and our place in it.

A PONDERABLE

Knowledge does not arrive fully formed. We have to work at it, taking it in turns to consider the evidence and offer our take on it. In hindsight even the most cutting-edge opinions can look utterly wrong, if not bizarre and laughable. But, our high-tech, inter-connected world is built on these ponderables, growing, changing step-by-step into an ever clearer picture of reality.

The story of the periodic table is one of classifying nature's substances. At first the material world was very much understood in terms of the mystical and magical realms. The elemental forms of the universe had religious and spiritual characteristics as well as physical ones. In this context it was just as sensible to control them with muttered chants as with heat or water.

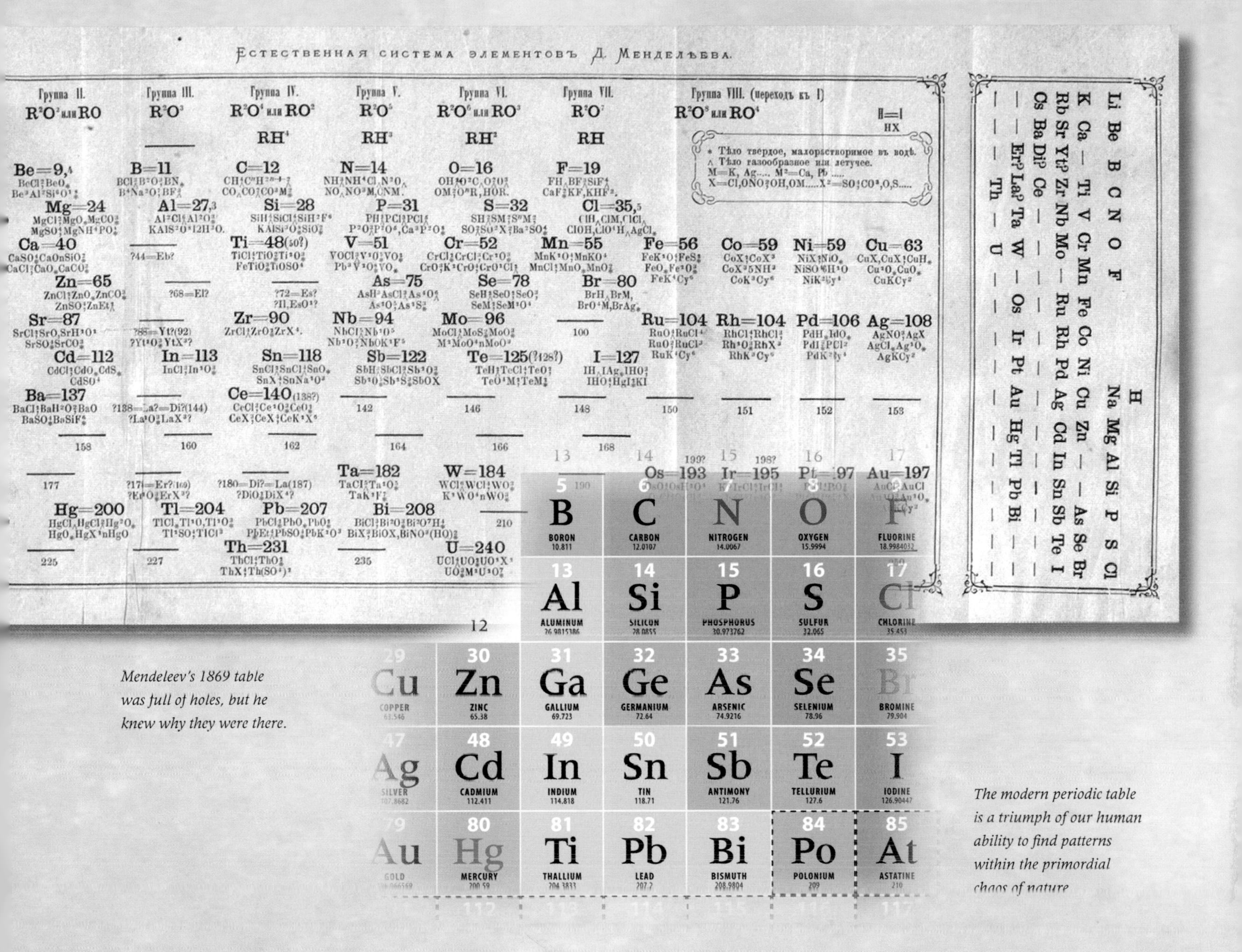

Mendeleev's 1869 table was full of holes, but he knew why they were there.

The modern periodic table is a triumph of our human ability to find patterns within the primordial chaos of nature

ANCIENT AND MODERN

In the ancient world, the elements could be counted on the fingers of one hand: water, air, fire, and earth. Nevertheless, we still saw definite patterns and links within this sparse set of substances. People began to manipulate substances—perhaps to make them more useful but definitely to make them more valuable—and a new breed of researcher called an alchemist showed that there was more to the elements than met the philosopher's eye.

An alchemist's workshop was a cornucopia of substances, frequently characterized as earths, oils, crystals, and airs. Over the years new techniques and apparatus were developed to combine, or fix, these substances, and gradually the mystical mumbo jumbo was stripped away to leave an empirical approach to investigation. This was the age of the scientist, who piece by piece began to reveal new elements—and debunk some old ones. By the 19th century, chlorine, uranium, helium … element after element was added to the list. At one point they were being discovered almost yearly and creating a headache for the catalogers. What was it that linked all these special substances but also made them all unique?

After several failed attempts, a table of elements that arranged them in a "periodic" system—in accordance with the repeating pattern of characteristics—was proposed by Dmitri Mendeleev. This periodic table of elements is still on the wall of chemistry labs the world over. It just worked, although Mendeleev and others were not quite sure why. Answering that question would require a good deal more pondering, and the story continues to this day.

UNDERSTANDING THE TABLE

The periodic table is the tool used by chemists to organize the known elements. The way the table is arranged allows a person to predict an element's properties at a glance, such as whether it is a metal or non-metal, whether it is reactive or inert and what other elements it is likely to interact with. Knowing a few simple rules is all that you need to settle those points. The elements are made from atoms, which in turn are made from heavy protons in a central nucleus, surrounded by light electrons. (A nucleus also normally contains neutrons which make it even heavier.) One element is different from another because it has a specific atomic structure, with a unique number of protons

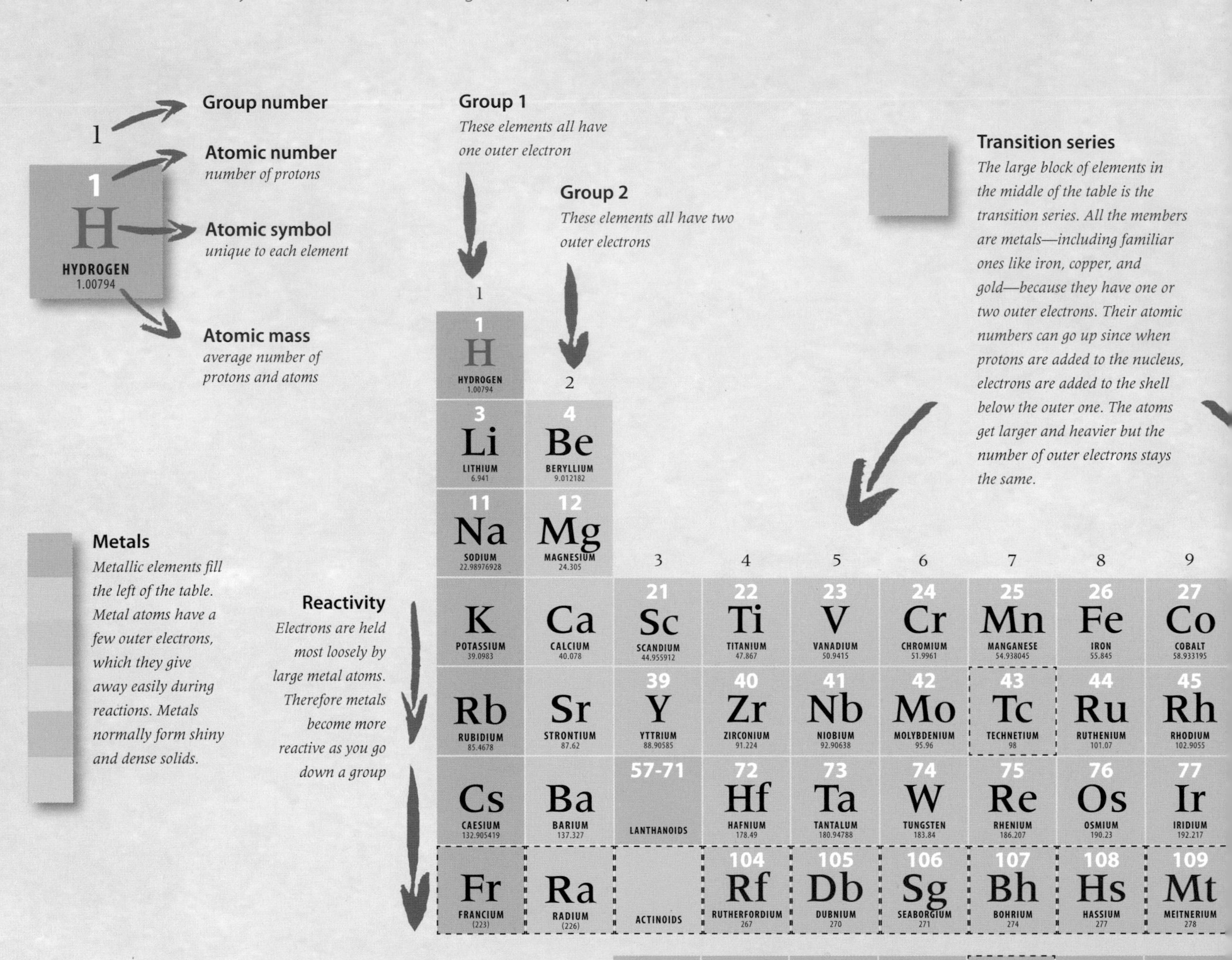

in the nucleus, and an equal number of electrons moving around it. The periodic table is ordered according to these atomic structures and begins in the top right with hydrogen, the lightest element with the simplest atom. With one proton orbited by one electron, hydrogen has the atomic number of 1. Moving left, helium has two protons and two electrons—and the atomic number of 2. The atomic number of lithium is 3, but the third electron is positioned a step out from the first two. Therefore it is placed in the table's second row, or period. The second layer, or atomic shell, has space for eight electrons, so the table continues to neon before starting a third period, and so on. Welcome to the periodic table.

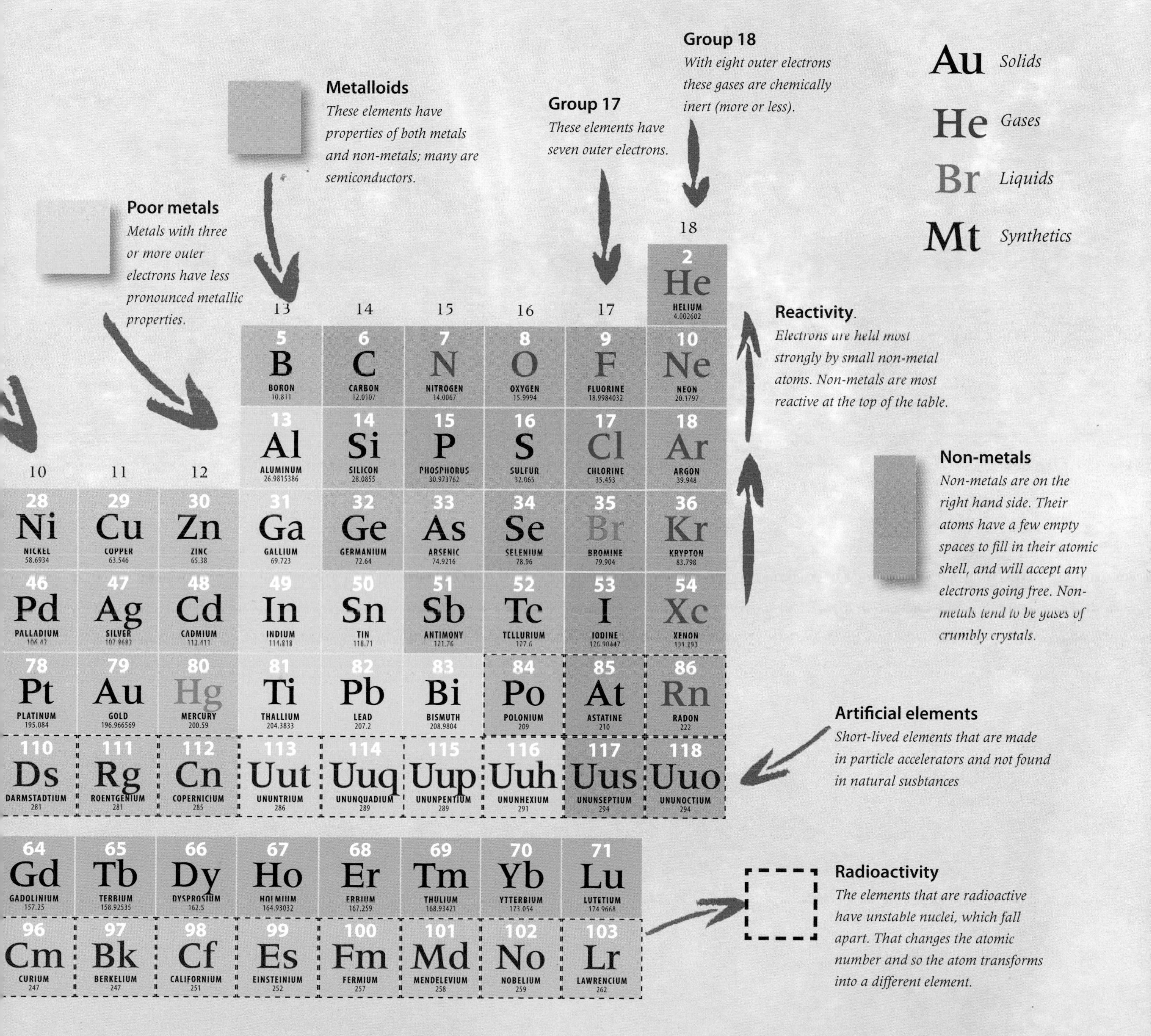

PREHISTORY TO THE FIRST CENTURY AD (50,000 BC – 0)

1 Stone Age Chemistry

AT ITS MOST BASIC, CHEMISTRY IS THE CLASSIFICATION OF MATERIALS, AND THAT PROCESS BEGAN AT THE DAWN OF HUMANITY. Fire, paints, and breadmaking are all examples of chemistry in action, and our earliest ancestors, even primitive hominids, made use of the chemical characteristics of the natural substances to hand.

Wood, sinew, and a stone arrowhead are combined into a bow and arrow in order to bring down a deer; and the scene is recorded for millennia using colored clays and a little imagination.

The Stone Age is the catch-all term for the earliest period of human culture, and actually began more than 2 million years ago, long before the modern human species *Homo sapiens* appeared on Earth. A milestone in evolution, it marks the development of simple tools by our direct ancestor *Homo habilis* (the "handy" man). The Stone Age is given that name because almost all of the evidence we have of human activity from this most distant period is made of rock, hewn by prehistoric hands. Nevertheless, it is

Flint hand axes like this were used from 1.8 million years ago by early humans for butchering animals and carving wood.

certain that stone was not the only material in use, especially by the time modern humans began to dominate in 50,000 BC. However, few of the implements fashioned from bone, antler, sinew, hide, and wood by our ancient forebears have come down to us intact. In eastern Asia, *Homo erectus* is thought to have developed a technology based on bamboo of which nothing survives.

Flames and food

There are several things that set humans apart from other animals, but the control of fire is one of the most significant. It is estimated that humans tamed fire as much as 1.8 million years ago, and there is little doubt that by around 120,000 BC lighting fires was within the scope of all African communities, if not those elsewhere.

Fire is the rapid release of energy when oxygen reacts with a fuel. The light of burning wood or dung freed humans from the schedule enforced by the rising and setting Sun, while the heat allowed them to modify the materials around them. Perhaps the best example of this is cooking, where the heat predigests food, breaking it down so its nutrients can be extracted by the gut easily.

Fire also drove early technology. Rudimentary vessels molded from wet clay were hardened by its heat, and these first pots would have held gathered grains, and any flour ground from them. Basic loaves baked on hot stones also rely on chemistry. Bread is not a simple mixture of flour and water; the two react to produce elastic glutens that are then kneaded together into the familiar springy dough.

Art and magic

Chemistry was also a factor in other ceremonial aspects of Stone Age society. One theory suggests that ancient art was a magical attempt to influence important events, such as the next hunt. Charcoal tips were the first pencils, while crushed clays were mixed with water to make primitive paint that was daubed or spat in patterns. The reds of cinnabar (mercury sulfide) and yellows and oranges of ochre (various iron oxides) make a recurring Stone Age color palette that is prevalent in the world's traditional painting styles to this day. Not for the last time, the world of science was shown the way by artists.

STONE AND STATUS

Embryonic cultures placed values on objects that went beyond their utility, although "value" would have begun with practicalities. A sharp antler was ideal for digging, harder wearing than wood but more flexible than stone. So, a good digging stick would have been jealously protected by its owner, more so than wooden items, which could be replaced more easily. The most prized object in a paleolithic tool box was the hand axe. This was a hand-sized, wedge-shaped stone that focused the force applied to the wide end into a blade-like edge at the other, just like a modern knife or axe works. Axes were frequently made from flint or other microcrystalline rocks, which fracture into hard, sharp-edged flakes. Ceremonial hand axes have been found that were too large to be of practical use, and only served to infer great status on their owners.

2 Naturally Pure

THE NATURAL WORLD IS A MIXED UP PLACE. AN EARLY HUMAN WOULD HAVE NEVER SEEN A REFINED PRODUCT; EVERYTHING AROUND THEM WAS MUDDLED AND MERGED TOGETHER. It is little wonder that nuggets of golden metal, objects that appeared wholly comprised of just one thing, attracted such a lot of attention. And they still do.

Several of the most abundant elements at the Earth's surface are metals, such as iron, aluminum, and calcium. However, these metals and most others are never found in a pure form. Instead, they occur as compounds, combined with silicon, oxygen, and other nonmetals. They form a plethora of minerals, natural compounds that make up the rocks, clays, and sands that give shape to the landscape.

However, here and there are glints of gold among the homogenous browns and grays. Gold is one of the few native, or naturally pure, elements. Native silver, copper, sulfur, and mercury can be found on rare occasions, but the chemistry of gold means that it very often occurs in a pure state. That characteristic, combined with its uniquely rich yellow hue, is what makes gold such a precious substance.

Copper is the most common native metal and as such was the first one to be worked on a large scale.

GOLD TODAY

All of the gold refined to date would form a cube with sides of 20 meters (more than 65 feet). Half of it is used to make jewelry, and ten percent has high-tech uses such as in electronics or in medicine. And of course gold is kept for its own sake. Forty percent of the total is stored in banks, bought and sold as investments; after all, the price has been going up for millennia.

Working metals

The first metalworkers used native sources, hammering them flat or molding the molten metal. Beads of native copper found in northern Iraq have been dated to 11,000 years ago. Gold is much rarer so fewer artefacts remain—the oldest come from Varna in Bulgaria, dating from 5,000 BC. The first gold mines appeared in Egypt 2,400 years later. Gold is a soft metal—biting a gold coin reveals its true worth—and is only suited for decoration. However, while other metals are weakened by corrosion or tarnished with age, people were quick to learn that the family gold did not fade as it was passed down the generations. Nor did it crumble into worthless rust. As a result, gold became a repository of wealth, a status it keeps to this day.

3 The Bronze Age

ANCIENT CRAFTSMEN WERE FAMILIAR WITH THE CONCEPT OF AN ALLOY, A MIXTURE OF TWO OR MORE METALS. GOLD WAS FREQUENTLY FOUND MIXED WITH THE LESSER METAL SILVER IN A NATURAL ALLOY CALLED ELECTRUM. However, it would be bronze, another alloy—this time man-made—that would change the world forever.

The Bronze Age is an indistinct historical period, a dimly remembered time when men could be gods that saw the erection of the Pyramids of Egypt, the Trojan Wars won using a wooden horse, and even the reported destruction of Atlantis—and it all started by accident.

In the 4th millennium BC, somewhere in Sumeria, now southern Iraq, metalworkers found that the charcoal fires they were using to heat native copper were producing more copper than was put in. They had stumbled upon smelting, where mineral earths mixed up with the pure copper were themselves rich in copper compounds. In a reaction with the burning charcoal, which is almost pure carbon, the minerals were reduced to pure copper. The same process also released tin from other ores, and the fortunate Sumerians found that the two metals if mixed when molten cooled into a solid alloy that was stiffer and tougher than either metal alone—bronze had been invented

Corinthian helmets such as this one, cast from a single piece of bronze, were worn by Greek soldiers for much of the 1st millennium BC.

Technological advantage

Bronze marked a huge step forward in human development. Stronger, longer-lasting tools made precision construction more widespread. Tough bronze plows could turn earth faster without breaking. Wheeled vehicles, including those used to transport ore from mines, were shaped and secured with bronze. And on the battlefield, a soldier armoured in bronze was safe from an enemy's copper weapons—but they were not safe from his bronze blade.

4 Using Iron

SMELTING WAS NOT ONLY USEFUL FOR REFINING COPPER AND TIN. OTHER ORES COULD BE REDUCED TO PURE METAL. "Reduced" was the term used because the metal that was produced weighed less than the original material. Metalworkers could identify different ores by their weight, texture, and even smell. Eventually they arrived at the ores of iron, a metal that is still the most widely used today.

Iron is almost never found in a native state, it is just too reactive. Nevertheless it is the most common metal element on Earth. Most of the planet's iron is far out of our reach in the hot dense core, but it is still a common ingredient in the rocks that make up the crust. Only oxygen, silicon, and aluminum are more plentiful in Earth's crust.

Metalworkers, seen here refining and molding iron, were specialist craftsmen. Thanks to the iron tools they created, agriculture became more efficient, resulting in a food surplus, which allowed Iron Age communities to concentrate on more than just survival.

Magical metal

However, the ancients were oblivious to the fact that iron was close at hand. To the ancient Egyptians, for example, iron was a magical "metal of heaven" that fell from the sky in meteorites. Since Egyptian copper and bronze were naturally high in arsenic impurities, which had a hardening effect, the ancient Egyptians did not feel the need to seek stronger, tougher alternatives. Instead, the drive for better materials came elsewhere in the ancient world, and that is where the Iron Age began.

CORROSION

The physical properties of strength and flexibility combined with its sheer abundance have made iron the most widely used metal. More than a trillion tonnes of iron is refined every year. However, it has one flaw—corrosion. Iron reacts slowly but surely with oxygen and water to form a flaky, porous mineral called goethite, more familiarly known as rust. Corrosion will one day turn all refined iron into red dust, although coatings and alloying can slow the process. Iron also expands as it rusts, so the steel reinforcing concrete will eventually cause it all to crack and collapse.

A once formidable steel dagger is transformed by corrosion.

Forging ahead

The first iron smelters appear to have been working just a few hundred years after the discovery of bronze in what is now northern Syria and southern Turkey. Iron works dating from 2000 BC have also been found in Tanzania, presumably the result of an independent discovery. By 1200 BC iron technology was being used from West Africa to the Caucasus, later to spread to China and western Europe.

A bronze smelter (with copper and tin ores being reduced together) requires a temperature of a little more than 1,000°C—well within the limit of a charcoal furnace. However, iron smelters work best above 1,500°C, which is beyond the reach of charcoal. This made early iron work a laborious business. In sub-optimal temperatures, the iron produced was a porous mass run through with impurities, or slag. This product known as bloom, or pig iron, needed to be forged—repeatedly hammered while being heated and cooled—to drive out the slag, and to consolidate the metal into a pure malleable form known as wrought iron.

Toughening up

Iron workers found that pig iron was too brittle but the wrought iron forged from it was too soft. However, the ingredient for a tougher form of iron that could replace bronze was already there in the furnace. Early iron furnaces, known as bloomeries, had large bellows to blow air through the burning mix of charcoal and iron oxide ore. The charcoal burned to produce carbon monoxide, which in turn reacted with the iron oxide, taking away its oxygens to become carbon dioxide, and reducing the ore to pure iron—albeit pig iron mixed up with other materials.

Wrought iron was soft because all the carbon in it was burned away in the forging process. However, if the forged metal was heated deep inside the charcoal and then dunked still red hot into water, the resulting metal was much tougher. This process, now known as carburizing, replaced the lost carbon as a toughened coating. In modern terminology, this mix of iron and carbon is called steel, even today a byword for durability and strength.

Ancient steel production was a laborious business, but the effort and skill was amply rewarded in the finished article. The Romans, the Han of China, the Vikings, and Japanese samurai, all owe their military success and resulting power to state-of-the-art steel technology.

BLAST FURNACES AND CONVERTERS

Bloomeries were refined through the ages. Waterwheels increased the air flow that fanned the smelter, while coke (a refined coal) was gradually introduced boosting the temperature within considerably. Limestone was also added as a flux, or cleaning agent that helped to remove impurities. As they grew in power, the humble bloomery was transformed into the mighty blast furnace. In 1855, Englishman Henry Bessemer developed a converter that turned pig iron directly into steel—and on a massive scale—without needing to forge wrought iron first.

The Bessemer Converter burned away impurities with a jet of hot air before a precise amount of carbon was added to produce steel.

5 Useful Minerals

METALS WERE NOT THE ONLY CHEMICAL SUBSTANCES REFINED AND PUT TO WORK BY EARLY CIVILIZATIONS. While highly valued gems have frequently survived the centuries, evidence of other mineral uses is patchy.

Fermenting alcohol was one of the first chemical processes exploited by early humans. Yeasts growing on stored fruits and grains would convert the sugars into ethanol naturally, but the Chinese led the way with a honey and rice wine deliberately made at least 9,000 years ago. Tanning leather also relied on biological action, with raw animal hides treated with soaked bark, dung, and even animal brains to convert its proteins into a durable, waterproof material that did not rot.

Much of what we know about ancient chemistry comes from analysis of pottery sherds for traces left by its contents. The production of pottery itself relies on a chemical process, with heat being used to change soft clay into solid earthenware.

Clay tablets were also the paper of the Bronze Age, and one written by a Sumerian physician around 2,100 BC records the substances he used regularly. The list includes sea salt (sodium chloride), soda ash (sodium carbonate) from burned plants, sal ammoniac (ammonium chloride) from coal ash, niter (potassium nitrate, later an ingredient in gunpowder), oils and fats, and alcohol (and perhaps vinegar), probably used as a solvent, antiseptic and anesthetic. However, the good doctor failed to record exactly what these items were for, so their medicinal efficacy remains shrouded in mystery.

The unmistakable colors of King Tutankhamun's death mask are produced by gold of the utmost purity inlaid with lapis lazuli. This blue stone is one of the first mass-produced gems and was mined in Afghanistan nearly 4,000 km away.

ARTIFICIAL ROCK

Concrete is a hard rock-like material made from granular material, often sharp sands, that are bonded together by a cement matrix. Its structure mirrors many sedimentary rocks, the difference being that when wet, a concrete slurry can be poured into molds, or forms, and left to set. The ancient Egyptians pioneered the use of concrete, using gypsum (calcium sulfate) and limes (calcium oxide) derived by heating limestones or shells. Rather than simply drying, the cement sets solid as its crystals absorb water molecules.

The Pantheon of Rome was built in 126 AD. To this day, its 43-m (141 ft) dome, made from a concrete cemented with volcanic ash, remains the world's largest to be made from unreinforced concrete.

6 Making Glass

THE RATHER MAGICAL TRANSPARENT PROPERTIES OF GLASS HAVE A SOMEWHAT MUNDANE PROVENANCE. GLASS IS A MYRIAD OF SAND GRAINS THAT HAVE BEEN FUSED BY INTENSE HEAT INTO A SINGLE LATTICE. It is formed naturally by violent events, such as lightning, volcanos, and meteorite strikes, but early humans also learned how to transform sand grains into objects of beauty.

The first civilization to make glass was probably ancient Egypt, around the middle of the 3rd millennium BC. Coppersmiths probably made glass by accident from sands mixed with their ores being smelted at high temperatures. Sand is generally comprised of small crystal fragments of silicon dioxide, otherwise known as silica or quartz, and the ancient Egyptians had access to an easy way of turning it into glass.

Nature makes glass by brute strength, applying temperatures way above the 1,700°C melting point of silica. The ancient Egyptian process involved mixing the silica with sodium carbonate, or soda ash, from lake beds near what is now Alexandria (the region is still a major producer of the material). When fired with the silica, the soda acted as a flux, a material that reduced the mixture's melting point significantly, making it possible to produce molten glass with a charcoal furnace.

The Egyptians mainly used glass as a beautiful glaze for their pots, and concerted glass production really began in Mesopotamia several centuries later. However, it took many more years before glass could replace pottery. Soda glass is slightly soluble and so is thinned and weakened by the water it holds. The solution arrived in 1300 BC, when quicklime (calcium oxide) replaced the soda ash flux.

Ancient glass was often blueish due to cobalt and copper impurities. Tin was added to produce white glass, while lead and antimony (another heavy metal) produce yellow objects. Some Assyrian glassware is red thanks to gold added to it, but how the ancient glassmakers achieved this is a mystery.

GLASS TOOLS

Obsidian is a black volcanic glass formed inside glutinous, slow-flowing lava. The Aztecs and Maya of Central America used obsidian for making tools. It could be flaked to produce razor-sharp edges, sharper than any stone blade. The glass was also hard enough to be used as chisels. Some suggest this glass-based technology was one reason why Mesoamerican civilizations did not develop advanced metal working.

Ancient glassware, such as this Egyptian bowl from the 4th century AD, was molded not blown.

7 The Classical Elements

MODERN CHEMISTRY IS SET AGAINST A BACKDROP OF THE ELEMENTS, A SET OF UNIQUE, PURE, AND INDIVISIBLE SUBSTANCES OF WHICH THE UNIVERSE, AT LEAST THE PARTS WE CAN SEE, IS COMPRISED. This concept, now rigorously tested by generations of scientists, began as little more than an intuition in the inquiring yet superstitious minds of ancient Greece.

Earth, air, water, and fire ... the four classical elements were not a Greek innovation. The Babylonians, Chinese, Egyptians and others recognized that nature's ingredients seemed to fall into simple groups that could be characterized as wet, dry, hot, hard, or soft, and so on. Frequently cultures linked the material world with the metaphysical, so elemental substances were also seen as manifestations of supernatural forces.

Not so the ancient Greeks. Many commentators suggest that Greece became the epicenter of science and philosophy simply because its pantheon of all-too-human gods, sitting squabbling atop Mount Olympus, provided such a poor set of answers to the big questions posed by human existence. Instead philosophers, by tradition led by Thales of Miletus (circa 600 BC), were left to puzzle out some solutions using nothing but observation, evidence, and logic.

The man who set out the basic four-element theory that permeated Western thought for the next 2,200 years was Empedocles, who lived in Sicily in the 5th century BC. He said the power of love strove to blend all elements, while the opposite power of strife forced them apart. The eternal battle between the two was what drove the constantly changing natural world.

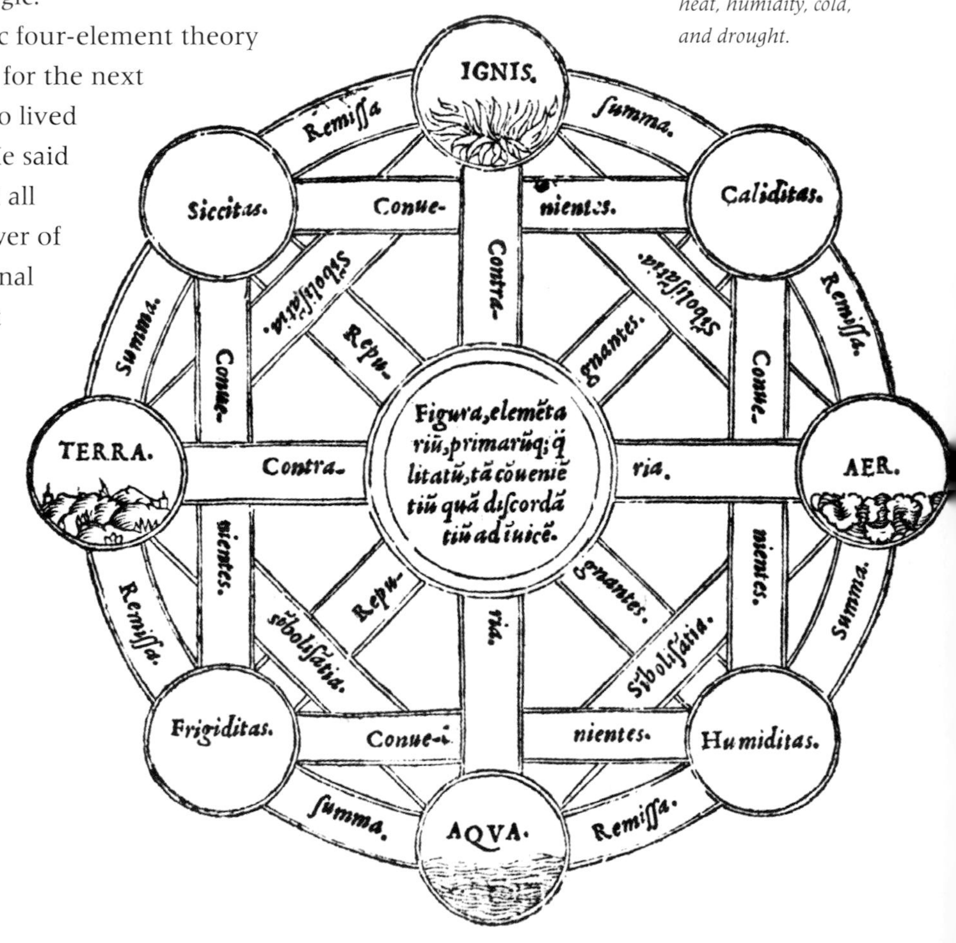

A 16th-century diagram shows the staying power of the classical elements. As well as fire, air, water, and earth, the network also shows their combinations—heat, humidity, cold, and drought.

THE FOUR HUMORS

The Greek physician Hippocrates, who is a founding figure of modern medicine, was a contemporary of Empedocles, and based his understanding of human physiology on four distinct body fluids, or humors. The humors were analogous to the elements: Black bile was earth, yellow bile was fire, phlegm was water, while blood was air. It was believed that if one humor began to dominate the others, ill health would result. Many early medical treatments, such as blood letting, were attempts to rebalance the humors.

8 Elektron and Magnesia

ELECTROMAGNETISM IS THE PHENOMENON THAT LIES AT THE HEART OF MODERN CHEMISTRY. It is also related to the release of light, is used in global navigation, and is one way a computer stores information. It may therefore be a shock to learn that our understanding of electromagnetism stems from a lump of amber and the son of Zeus.

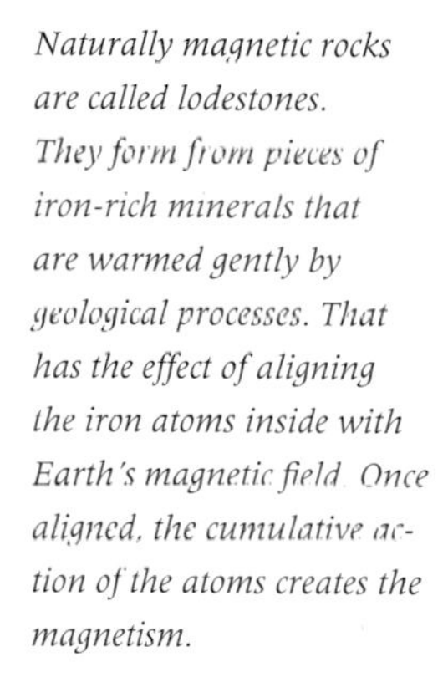

Naturally magnetic rocks are called lodestones. They form from pieces of iron-rich minerals that are warmed gently by geological processes. That has the effect of aligning the iron atoms inside with Earth's magnetic field. Once aligned, the cumulative action of the atoms creates the magnetism.

The Greek word for amber is *elektron*, which is the root of modern words such as electricity and electron. Amber is a fossilized dollop of tree resin, and to the Greeks its name was an allusion to how the clear orange stones appeared to trap sunlight. In the 4th century BC, the Greek philosopher Theophrastus wrote a compendium of rocks and their properties. Within was one of the earliest records of amber, listing it as being unusual in that it was able to attract lightweight objects, such as feathers or dust. Rubbing a sample of amber gives it a small static electric charge that manifests itself as an attractive force, just like a child's balloon clings to a sweater or pulls long hairs away from the head. Theophrastus gives no explanation—that was 2,000 years away—but it is this reference to amber that put the *electro-* into electromagnetism.

FINDING THE WAY

At the same time that Greek philosophers were studying lodestones, Indian surgeons were using them to clean arrow wounds of iron flakes, while the Chinese made the first compasses from free-floating lodestones. The compass only became a navigational tool in the 11th century AD: before then it was used in *feng shui* and fortune-telling.

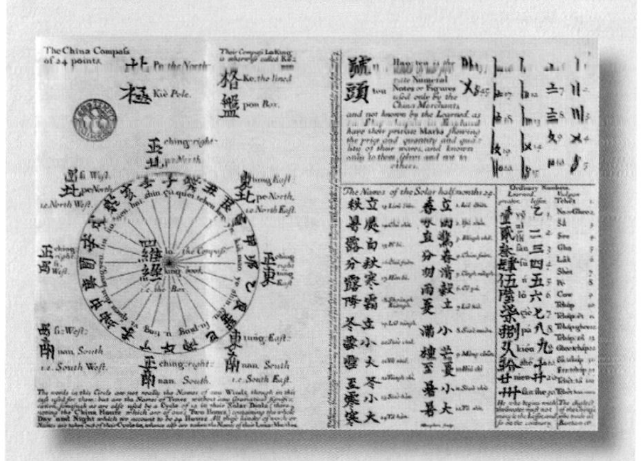

An 18th-century diagram of a Chinese navigational compass. A magnetic needle floated on water at the center.

Rocks of Magnesia

Theophrastus also mentions that *magnítis líthos*—the stones of Magnesia—not only attracted but also repelled each other in equal measure. He was referring to magnetite, an oxide of iron that is named after the Magnesia region of central Greece, the mythical kingdom of Magnes, son of Zeus. (Magnesia is rich in minerals and also lends its name to magnesium and manganese.) The intuitive link between magnets and electricity was not confirmed until the 19th century, but has since transformed our understanding of matter.

9 Atomism

THE WORD *ATOMIC* IS ALMOST SYNONYMOUS WITH *MODERN* SO IT MAY COME AS A SURPRISE THAT THE THEORY OF ATOMS IS A THOROUGHLY ANCIENT ONE. Its greatest exponent was Democritus, a Greek philosopher who lived 2,400 years ago. He saw matter as a set of indivisible bodies moving in a universal void.

Despite this rather maudlin likeness, Democritus is remembered as the Laughing Philosopher, a cheerful soul who saw no goal or purpose to the universe.

Atomism, the doctrine proposed by Democritus, was a response to questions over how nature was able to be constantly changing but retain its properties. Some of his forebears had suggested that change was merely illusory. For material to move, they said, it must travel into a place where there was nothing—and how could nothing change into a "thing?" And how could dividing matter result in a "nothing" occupying its place?

For Democritus, in line with the teachings of his mentor, Leucippus of Miletus, the answer was simple. Matter could not be divided indefinitely. Instead all things were constructed of minute, indivisible solids, termed *átomos*, meaning "uncutable." Any changes in nature were merely due to atoms being rearranged. Democritus reasoned that atoms need not be identical but had characteristics that could explain the wealth of substances that he observed in nature: sticky or hooked atoms clustered into solids, while smooth ones flowed past each other in water and wind.

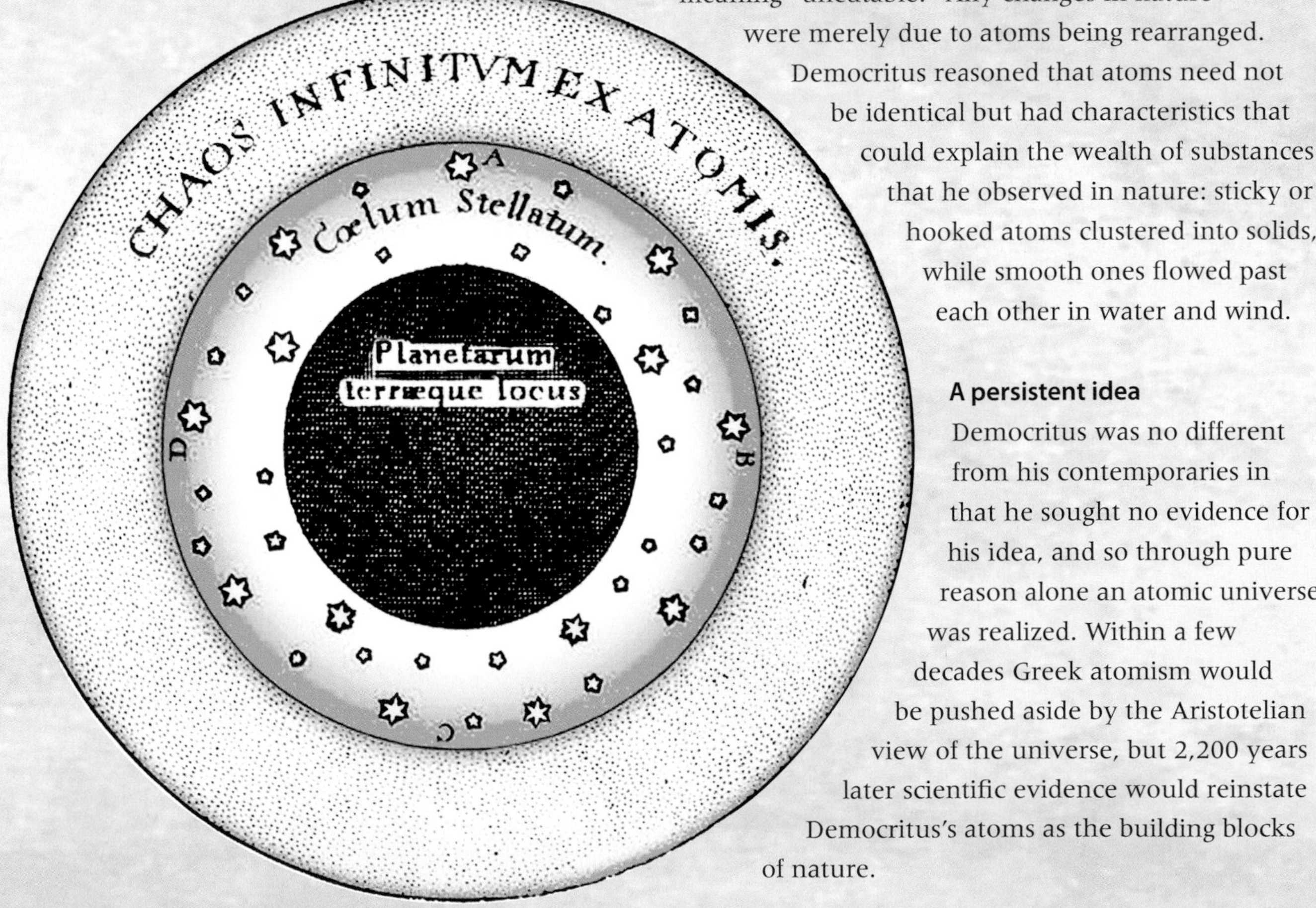

Democritus did not get everything right: the Democritean universe had three concentric regions with the planets at the center surrounded by the heavens, and all encompassed by an infinite chaos filled with atoms.

A persistent idea

Democritus was no different from his contemporaries in that he sought no evidence for his idea, and so through pure reason alone an atomic universe was realized. Within a few decades Greek atomism would be pushed aside by the Aristotelian view of the universe, but 2,200 years later scientific evidence would reinstate Democritus's atoms as the building blocks of nature.

10 Platonic Solids

PLATO WAS A CONTEMPORARY OF DEMOCRITUS, BUT BY NO MEANS A SUPPORTER. He rejected the chaos of atomism, even advocating that Democritus' books be burned.

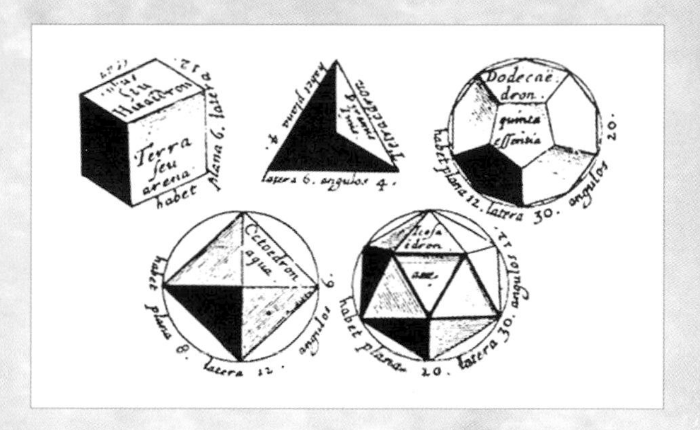

To Plato, like Pythagoras before him, the only unchangeable thing in nature was mathematics. The elements had to conform to the order—and beauty—of numbers. Plato's elements took the form of the regular polyhedrons—the so-called Platonic solids—of which there can only ever be five. Earth was formed of cubes, corpuscles of fire were tetrahedra, air was made of octohedra, and water took the form of the 20-sided icosohedron. That left one more. Plato suggested that the space between the elements was filled with the dodecahedra.

Plato attributed each solid to an element according to how their shapes related to each element's characteristics.

11 Buddhist Atomism

THE IDEA OF ATOMS WAS NOT RESTRICTED TO WESTERN THOUGHT. In the 4th century BC, THE Buddhist philosophy of India also described the elements in this way.

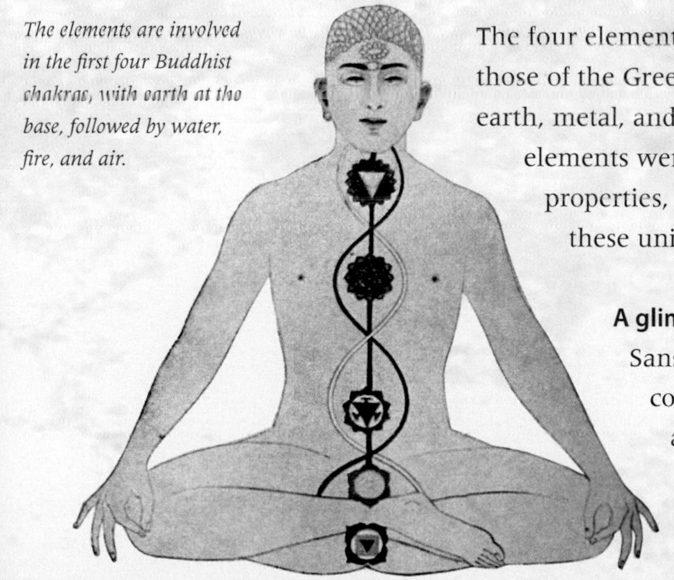

The elements are involved in the first four Buddhist chakras, with earth at the base, followed by water, fire, and air.

The four elements of the Buddhist world view were the same as those of the Greeks. (However, the Chinese had five: fire, water, earth, metal, and wood.) In ancient India, Buddhists believed that the elements were composed of *paramânu,* basic units with a variety of properties, such as motion or solidity. In different combinations these units produced nature in all its forms.

A glimpse of the future

Sanskrit Buddhism differentiated single *paramânu* from combinations, which were termed *samghata paramânu* and were analogous to what we understand today as molecules. In later developments, Buddhist atomism referred to subunits that exist only as atoms—an early suggestion of subatomic particles.

CHAKRAS

According to several Eastern philosophies, the chakras are energy centers located throughout the body. The word means "wheel" in Sanskrit, and a chakra forms the connection between the physical and spiritual realms.

12 Ether: Aristotle's Fifth Element

ARISTOTLE, PLATO'S STAR PUPIL, WAS ONE OF THE MOST INFLUENTIAL THINKERS OF ALL TIME. His ideas about the Universe held sway for nearly 2,000 years. At the heart of his theories were not four, but five elements.

While attending the Akademia, Plato's school in an olive grove beyond the city walls of Athens in the mid-4th century BC, Aristotle would have heard his teacher talk of ether—a substance that filled the space between the elements. It was always there, Plato said, unseen but necessary to delineate material from the void. As Aristotle rose to supersede his master, he overhauled this idea, and promoted ether to being the quintessence, or fifth element.

Atlas, the Greek god tasked with supporting the heavens, holds up Aristotle's Universe, which although undoubtedly heavy was considerably less massive than the cosmos as we currently understand it.

Combining ideas

Through modern eyes it is baffling that Aristotle, the man credited with explaining the Universe to humanity for centuries, sought no systematic proofs for his ideas. However, his theories were based on observations of natural materials and related phenomena. He used the forms and characteristics of the four classical elements—earth, water, air, and fire—as the foundations for constructing a fully functional mental model of the Universe, with Earth and humanity positioned securely at its very center.

Aristotle believed that the four earthly elements combined in various proportions to form the many substances he saw in nature. Heat, aridity, cold, and damp were all evidence of the presence of these substances. The smoke from smoldering wood was the air escaping from within, the resin forced out by the heat was the water, while the ash left behind was the earthen component. And of course the flames were its fire. Flowing lava was water, fire, and earth combined, while a flint let off sparks as the lightweight fire in it tried to escape from the heavy earth.

DISPROVING ETHER

How does light pass through a vacuum? One answer was ether, a medium that existed even where there was nothing. The Michelson–Morley experiment of 1887 was designed to show that light beams were slowed slightly as Earth moved through the ether. However, the experiment showed no effect: ether was dead.

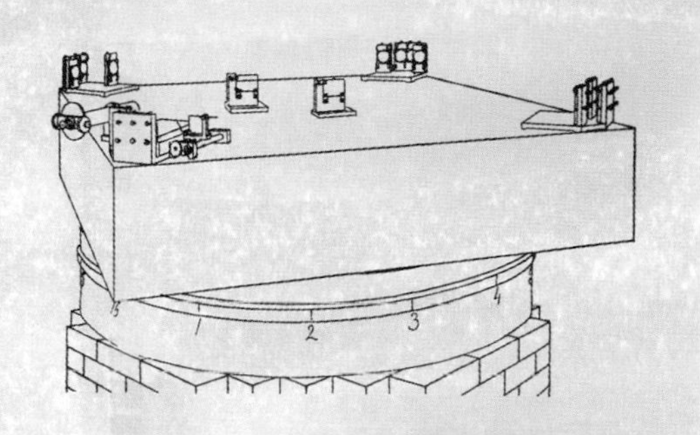

The famous failed experiment split and recombined light beams to detect any speed differences.

Pure layers

Aristotle reasoned that the driving force behind natural events was the desire of elements to separate into their pure forms. Earth was the most basic of elements and the heaviest, so it sunk beneath the rest forming the land and seabed. Water formed the next layer, followed by air and then fire. The tumultuous activity of Earth such as volcanic eruptions or earthquakes, and the torrential downpours of storms, provided further evidence of the elements finding their ways to their rightful positions.

As such the elements formed into four rings which formed the terrestrial, mortal realm in Aristotle's Universe. This reached out as far as the Moon, while beyond lay the Sun and the five known planets which moved in ever increasing circles around Earth. Everything was enclosed by a final crystal sphere of stars. The heavenly region of the Universe beyond the Moon was where Aristotle's ether was located. It was beyond the reach and experience of humans and did not mix with other more lowly elements. This was evidenced, according to Aristotle at least, by the unchanging nature of the heavens. The word ether is derived from the Greek for clarity and purity. In turn the English word *quintessential*, meaning perfect, is derived from ether's fifth and most pure position (from the Latin *quintus*) among the classical elements. Amazingly, it took Einstein's theory of special relativity in 1905 to explain how the Universe could work without an invisible, all-pervading ether.

REVEALED TRUTH

Aristotle's theories predated Christendom, and were adopted by church leaders as revealed truth—a look under the hood at the workings of Creation. Aristotle's Universe provided answers to questions not covered in the Bible, and did little to contradict orthodox teachings. However as evidence against Aristotle's theories began to accumulate in the 16th century onwards, the Church found itself at odds with many great scientists, who were actually challenging Greek philosophy rather than the Christian message. In 1991 the Vatican finally renounced Aristotle's teachings of an Earth-centered Universe.

Aristotle is at the center (in blue) of Raphael's fresco School of Athens *(1511), which decorates the Papal Palace in the Vatican City.*

THE DARK AGES TO THE MIDDLE AGES (0 – 1600)

13 Black Magic, the Birth of Alchemy

THE SCIENCE OF CHEMISTRY BEGAN AS THE PRACTICE OF ALCHEMY. An alchemist was part doctor, part inventor, and part magician, but nevertheless began the in-depth investigation of the stuff of nature.

Alchemy was not invented, it just emerged from the shadows. The word is an Arabic version of the Greek term *chemia*—the *al-* is "the" in Arabic. The root of chemia is uncertain. It may be a corruption of the Greek word for mixture, which would be apt, but most authorities suggest that the word actually refers to the land of Egypt. The ancient Egyptians referred to their fertile land as *Khmi,* meaning "black earth," and so let us begin the story of this murky practice in Alexandria at the mouth of the Nile.

As this 17th-century Dutch painting shows, alchemy was very much a practical occupation, where trial and error superseded theory.

MERLIN, A WIZARD

Alchemists are enshrined in Western culture as wizards and magicians. One of the most famous was Merlin, King Arthur's mentor.

Alexandria was founded and named for Alexander the Great, Aristotle's most famous pupil, who also happened to conquer most of the known world while still in his twenties. The great port city took over from Athens as the seat of learning in the latter part of the Classical era (up to circa 300 AD).

Few of the city's alchemists have been remembered by name, but we know that they had differing goals. Some were artisans, skilled in metal work, while others were apothecaries, preparing curative potions. Yet more were wild mystics, who were searching for ways to take control of the four elements as described by Aristotle—and perhaps grow rich and powerful in the process. Alexandria's alchemists were undoubtedly influenced by metaphysics, such as Greek astrology, but also the teachings from China and Persia.

Whatever their motivation, generations of alchemists developed useful techniques and apparatus. They learned how to distil liquids and assay (test) their purity, to sublimate vapors, to gild, and to use dyes. All these skills and more would one day be used by scientists to reveal the true nature of the elements.

14 Secret Knowledge: Jabir's Gibberish

KNOWLEDGE IS POWER, AND ALCHEMY WAS AS HIGH-TECH AS THE DARK AGES GOT. One Islamic alchemist was so good at hiding his discoveries that his writings needed a new word to describe them.

Jabir is often referred to in European texts by the Latinized name Geber.

As Europe fell into the Dark Ages after the fall of the Roman Empire in the 5th century AD, centers of scholarship moved east to the expanding Islamic empire. Jabir ibn Hayyan was a leading Persian alchemist in the 8th century. Jabir followed Aristotelian teaching that the elements could change from one to another, and formed the mercury–sulfur theory. Sulfur was earth that was transmuting into fire, while mercury was water on the verge of becoming air. Metals, he believed, were mixtures of sulfur and mercury, and therefore changing one (say copper) into another (perhaps gold) merely involved altering the proportions of these constituents. Perhaps to hide these findings from the uninitiated, or perhaps because he believed it was a central requirement, Jabir's records were written in a dense mystical argot that gave rise to a new word—gibberish!

15 Practical Magic

ALTHOUGH IT NEVER LOST ITS LINK TO THE SUPERNATURAL, ALCHEMY DID MUCH TO FURTHER HUMAN CIVILIZATION, FOR GOOD AND BAD. We can thank alchemists for developing the technology behind the perfume industry, fine china, and even how to make chocolate sauce.

German alchemist Berthold Schwarz is said to have invented gunpowder independently in the 14th century.

Many of alchemy's greatest contributions were the unintended consequences of a quite unrelated endeavor. The 9th-century Chinese inventors of gunpowder were not looking for an explosive. They were attempting to combine the putative health-giving qualities of niter with sulfur and herbs to create a warming medicine. And the result was a mixture with enough explosive power to change the way wars were fought forever.

The legacy of Mary the Jewess, a little-known alchemist from the Alexandrian era, has brought more pleasure. She is known only through references to her in the writings of others and could have lived any time between 100 and 400 AD. Her lasting contribution was a water bath apparatus that could heat volatile solids gently and evenly so they melted rather than burned. The *bain-marie* (French for "Mary's bath") is used today mostly for melting chocolate or preparing crème caramel.

With a name like Thrice Great Hermes, one might expect this literary figure to have left more of a mark. More than likely a composite of pre-Christian scholarship and prophecy than a real person, the mystical writings attributed to Hermes Trismegistus (his Latin name), known as the Hermetic Corpus, occupied the minds of many an alchemist from Cyrene to Khorasan for centuries. However, the only lasting contribution from it is the concept of a hermetic seal. This referred initially to a secret method of sealing glassware, probably involving wax and no doubt a few muttered spells. Today's hermetic seals, high-tech or otherwise, can include anything from the door to a semiconductor clean room to a tight lid on a pickle jar.

THREE POWDERS IN ONE

The constituents of gunpowder are saltpeter (or niter), the common name for potassium nitrate; charcoal, which is more or less pure carbon; and sulfur. All three are pulverized and thoroughly mixed. The nitrate provides extra oxygens so the powder burns very fast, releasing high-pressure gases with a bang.

Saltpeter (top), charcoal, and sulfur in the required proportions for gunpowder.

The influence of Islamic alchemy

The teachings of Islam give special import to learning. This is perhaps why a trend developed in Arabic alchemy to reduce the mystical mumbo jumbo and focus on recording the process as well as the result.

In the 9th century, al-Razi, a wealthy, educated man from the hills above Tehran, did more than most to propel alchemy towards modernity. He took the knowledge of the earlier Alexandrians, especially that of Zosimos, and extended it in the light of the wide range of materials available to him from the immense Arab empire. He classified the minerals as bodies (malleable metals); salts; vitriols (containing sulfur); spirits (mercury and sulfur and other things that vaporize easily); boraxes; and stones (generally brittle rock-forming minerals).

Arabic legacy

Several words used in chemistry have come down to us from Islamic alchemy. Talc, realgar, cinnebar, and arsenic are all names derived from Arabic or Persian. al-Razi himself reported two of the most familiar loan words—alcohol and alkali. Alcohol is derived from kuhl, a dark spirit (in al-Razi's terminology) used in traditional eye makeup. When ethanol was distilled from wine, it too was identified as the spirit, the essence, or *al kuhl* of the original. Refined alcohol was used to dissolve herbal oils, making Muslim cities like Cairo and, later, Istanbul the center of the world perfume trade. Alkali comes from *al qaliy,* the Arabic for ash mixed with lime (calcium hydroxide). al-Razi used the word when describing how to turn potash (potassium carbonate) into a caustic substance known as "sharp water" (potassium hydroxide) that could even dissolve stones.

The blue tiles that are so typical of Islamic art were made with cobalt salts mined in Persia about 1,000 years ago. The dye was traded far and wide; it was used in fine Chinese porcelain and in the luxury blue glass of Europe.

16 A New Approach

HARRIED BY ENEMIES ON ALL SIDES, THE ISLAMIC WORLD, SO RICH IN SCHOLARSHIP, BEGAN TO FRAGMENT AND LOSE INFLUENCE. By the 10th century, inquiring minds in Europe were also investigating the chemical nature of substances.

A 14th-century illustration outlines the process of distillation, where liquid mixtures are separated into pure samples. To extract alcohol from water, for example, the mixture is heated but not boiled. Any vapors given off are collected in separate vessels where they condense into pure alcohol.

The works of Jabir, al-Razi, and others were brought back from the Holy Land by crusaders or were captured in the reconquest of Iberia. Eventually they found their way to Europe's monasteries, where most literate intellectuals were based at the time.

Skeptical views

The great wealth of knowledge that was accruing in Europe's monasteries and embryonic universities resulted in a new type of philosophy known as scholasticism. This ranked knowledge according to the perceived authority of its source, with the works of Aristotle second only in importance to the Bible. The most influential scholastic was a German friar called Albertus Magnus, who was the leading Aristotle expert of the age. The 1270 condemnation by church leaders in Rome of much of Aristotle's natural philosophy, including his theories of elements, resulted in Albertus' commentaries on it becoming more popular than ever. Like the Tehran-born philosopher al-Kindi before him, Albertus Magnus was nevertheless skeptical of whether one element could be transmuted into another, as predicted by Aristotle's theories. He had doubts in general about the results reported by alchemists, although that did not stop him from making a few claims himself. He recorded his research into oil of fortis—

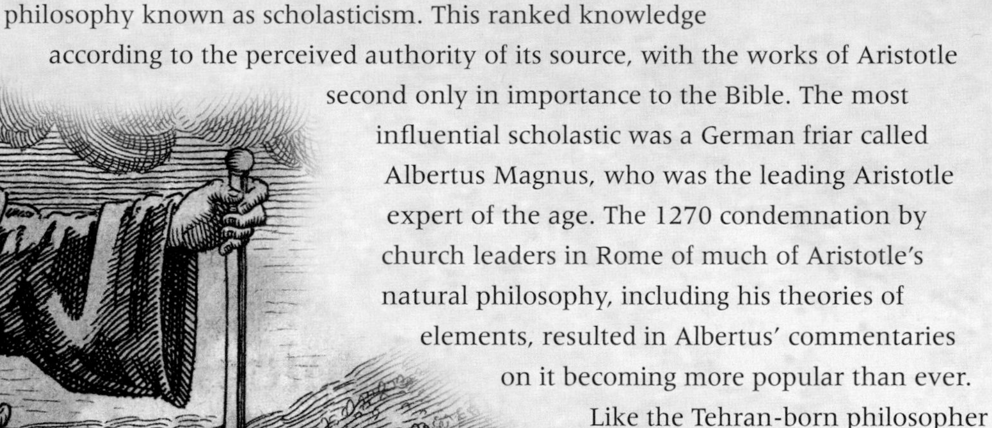

Roger Bacon is remembered as one of the first true scientists, and after his death became known as Doctor Mirabilis, a name that would not be out of place in a comic book but that referred to the wonderful teachings that Bacon left for future generations.

known as nitric acid now—a liquid so potent it could dissolve silver. Albertus said that the resulting liquid turned his skin black. (It contained silver nitrate, the light-sensitive chemical used in early photographic films.)

Material advance

Roger Bacon was an English contemporary of Albertus. Initially he followed the same scholastic approach and was a vocal supporter of the German's work. However by the 1250s he changed his mind and questioned why the authority of a theory took precedence over empirical evidence. In his own words: "For if any man who never saw fire proved by satisfactory arguments that fire burns. His hearer's mind would never be satisfied, nor would he avoid the fire until he put his hand in it that he might learn by experiment what argument taught."

In so doing Bacon set Europe's alchemists on the road towards scientific rigor, although there was still a long way to go.

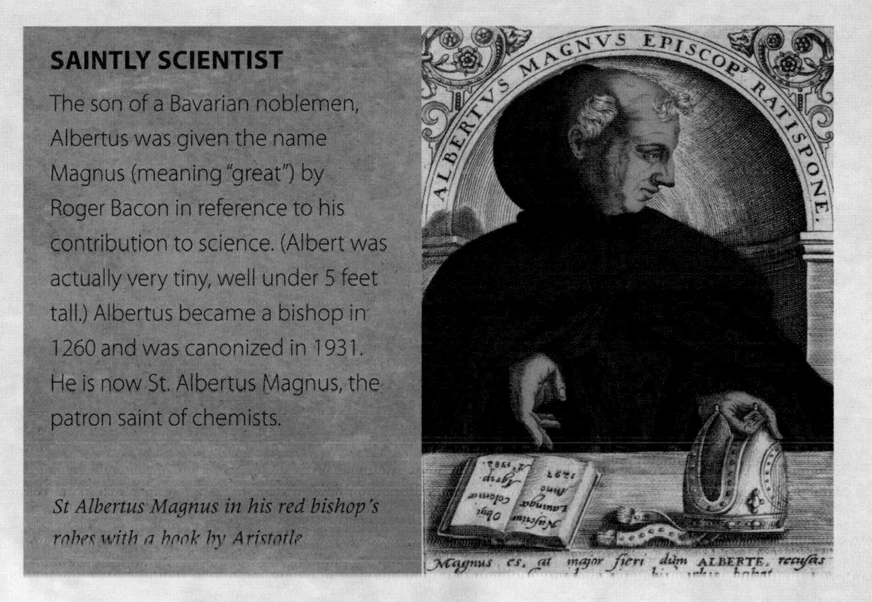

SAINTLY SCIENTIST

The son of a Bavarian noblemen, Albertus was given the name Magnus (meaning "great") by Roger Bacon in reference to his contribution to science. (Albert was actually very tiny, well under 5 feet tall.) Albertus became a bishop in 1260 and was canonized in 1931. He is now St. Albertus Magnus, the patron saint of chemists.

St Albertus Magnus in his red bishop's robes with a book by Aristotle

17 The Litmus Test

THE STUDY OF ACIDS FORMED THE FRONTIER OF ALCHEMY IN MEDIEVAL EUROPE. Not even gold was impervious to these potent materials, and in 1300, a new test of their strength was discovered.

Weak organic acids, such as those in vinegar and citrus juices, had been known for millennia. However, by the 13th century, altogether more potent acids derived from minerals were being rediscovered by European alchemists. Of most interest were oil of vitriol (now known as sulfuric acid) and oil of fortis (nitric acid), which when mixed formed *aqua regia*, or royal water. This mixture had the magical property (for the time) of being able to dissolve gold. Could this be the key to unlocking gold from everyday materials? In 1300 the Catalan alchemist Arnaldus de Villa Nova created a useful new way of testing for the presence and strength of acids. He discovered that a purple dye made from a lichen turned red when it was added to an acid—growing darker with the potency of the acid. The same dye turns blue when in an alkali, the opposite of an acid. This became litmus, the first acid-base indicator.

18 Wizards and Witchcraft

Alchemists busy themselves collecting aqua vitae, the water of life that could cure all ills and even make drinkers immortal. It is likely that these alchemical potions had a high spirit (or alcohol) content, and even today local schnapps and other distilled drinks are often referred to as an aqua vitae.

MEDIEVAL EUROPE WAS A HARSH PLACE TO LIVE WITH MOST PEOPLE FACING CONSTANT MISERY AND THE POSSIBILITY OF AN EARLY, SUDDEN DEATH. The tide of public opinion turned against the alchemists.

At its very core, alchemy was about putting Aristotle's as yet unquestioned teachings into practice, and finding a way to transmute the elements. The underlying motivation for that soon became to transform inexpensive "base" materials into valuable gold and silver. Research in this field progressed for almost 2,000 years and was the direct ancestor of modern chemistry. However, during the long transition between the two during Europe's Middle Ages the work of alchemists was increasingly viewed with skepticism, and even with fear and hostility.

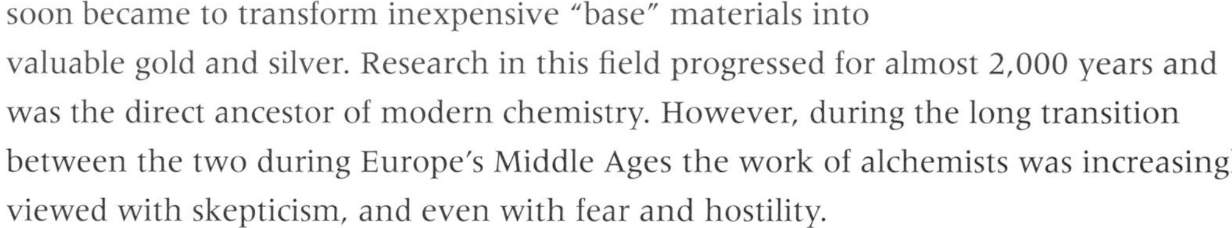

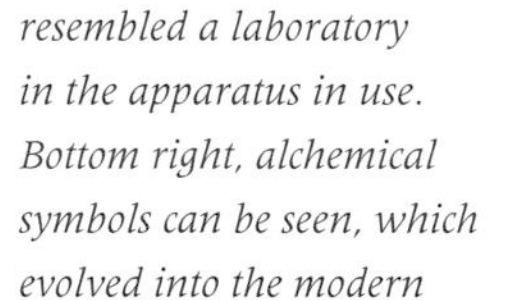

Miracle workers

By the turn of the second millennium, most alchemists thought that the key to success was to find what their Arab forebears called *al iksir*—a miraculous substance that turned base metals into gold. Those who thought this material was solid began to refer to it as the "philosopher's stone," while if liquid it was termed the "elixir." Swindlers promising to convert small fortunes into larger ones gave alchemists a bad name.

In addition to making its owner fabulously wealthy, the miraculous stone or spirit (the elixir was most probably volatile stuff), alchemists also contended that it could provide immortality and eternal youth. Hence the term elixir of life has come down to us, along with *aqua vitae* (water of life) and panacea (a cure-all named after a Greek goddess).

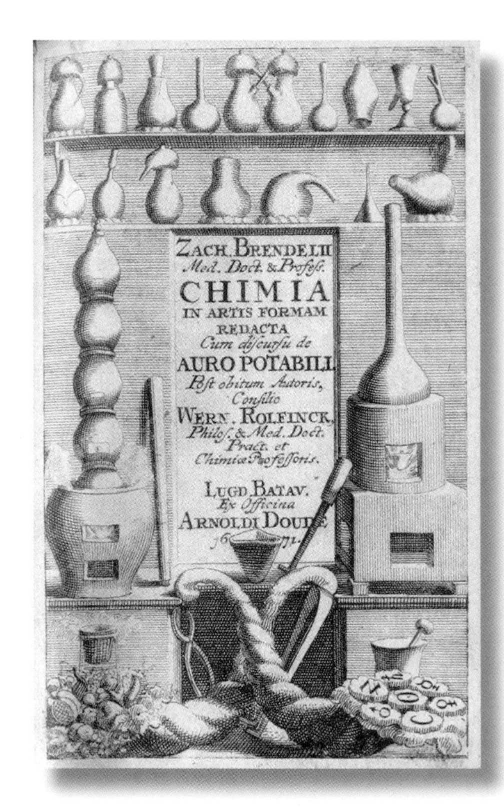

ZACH. BRENDELII
Med. Doct. & Profess.
CHIMIA
IN ARTIS FORMAM
REDACTA
Cum discursu de
AURO POTABILI.
Post obitum Autoris,
Consilio
WERN. ROLFINCK.
Philos. & Med. Doct.
Pract. et
Chimiae Professoris.
LUGD. BATAV.
Ex Officina
ARNOLDI DOUD

An alchemist's workshop resembled a laboratory in the apparatus in use. Bottom right, alchemical symbols can be seen, which evolved into the modern chemical shorthand.

Miracles required

Life was short and brutal in the Middle Ages. In the mid-14th century, the Black Death killed at least a third of the population of Europe. In response, some alchemists offered the prospect of a cure—an unbearable temptation for the desperate and the dying. But no doubt some alchemical potions only

ALCHEMY OUTLAWED

England's Henry IV banned alchemy in the Act Against the Multipliers of 1404. (Multiplication was the process of refining an elixir or philosopher's stone.) The act was meant to protect against swindlers, while ensuring that no one could gain enough wealth to rise up against him. His grandson failed to fend off civil war and offered temporary licences to alchemists, perhaps to gain an advantage, but alchemy remained banned for 250 years.

added to the suffering, and of course none were effective, ruining the reputations of alchemists. Ordinary people were unaware of the criticisms leveled at alchemy by the likes of Roger Bacon and Albertus Magnus, but the clergy would have told them that such practices were against God. So alchemy's occult practices, so often fused with local pagan traditions, gradually became associated with the devil. Alchemists became witches and warlocks to be feared, exposed, and destroyed.

19 The Nature of Metals

WITH ALCHEMY REDUCED TO SUPERSTITION AND INFAMY, A MORE PRACTICAL APPROACH WAS NEEDED. A German physician set the standard.

With the get-rich-quick message of alchemy consigned to history, the emphasis shifted to expanding the collection of useful minerals, especially metals. In 1556, Georg Pawer, a physician from a mining town in what is now the Czech Republic, published a compendium of mineralogy, called *De Re Metallica* (The Nature of Metals). He chose the nom de plume of Agricola (meaning farmer in Latin, as pawer does in German). The book outlines how to recognise ores, where to find deposits, and the latest technology of mining and smelting as it stood in the 16th century. Agricola's was not the only technology manual of the era but it proved the most significant with copies still in use 200 years later.

De Re Metallica *gives advice on how to use water-wheels to power the operations of a mine, such as rasing spoil from the shaft or working bellows in the smelter.*

THE ENLIGHTENMENT (1600–1800)

20 Mapping Magnets

IN 1543, NICOLAUS COPERNICUS HAD PLACED EARTH IN ORBIT AROUND THE SUN AND SHATTERED ARISTOTLE'S UNIVERSE. Scientists began to reform the puzzle, piecing it together this time by looking at the evidence. William Gilbert turned his attention to the Earth itself.

The full title of Gilbert's 1600 magnum opus was De Magnete, Magneticisque Corporibus, et de Magno Magnete Tellure *(On the Magnet and Magnetic Bodies, and on the Great Magnet the Earth). It contained illustrations of his experiments with a magnetic terrella, or scale model of Earth.*

Like many of the people that populate the story of elements, English scientist William Gilbert had a varied career. He was a university administrator, an astronomer and the personal physician of Queen Elizabeth I. However, he is remembered as the father of electricity. That word itself is said to have been coined by him, in a reference to any phenomena that resembled the attraction behavior of amber (*elektron* in Greek).

On magnets

However, Gilbert's *magnum opus* (major work) was *De Magnete*,(On the Magnet) published in 1600. In it he reveals that our whole planet is a magnet. Just as the

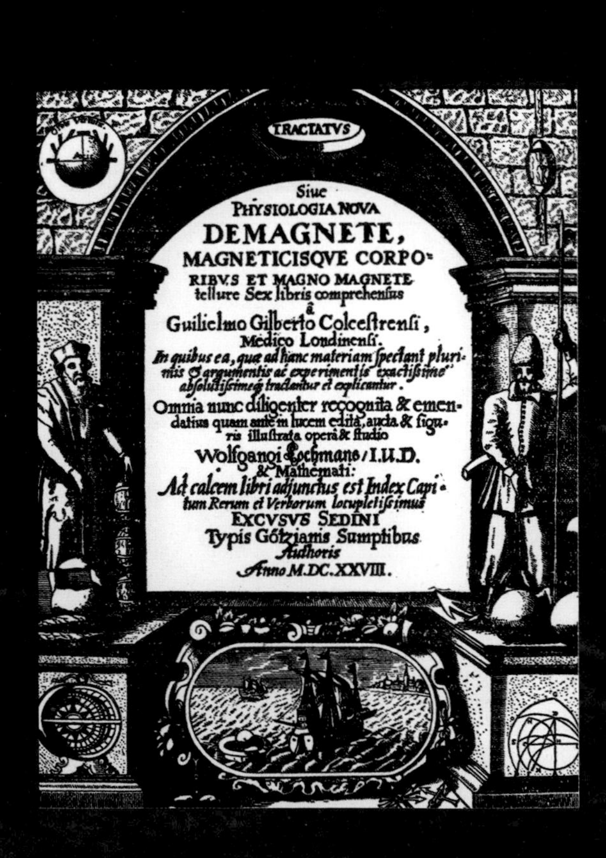

TRACTATUS
Siue
PHYSIOLOGIA NOVA
DE MAGNETE,
MAGNETICISQVE CORPO-
RIBVS ET MAGNO MAGNETE
tellure Sex libris comprehensus
à
Guilielmo Gilberto Colcestrensi,
Medico Londinensi.
*In quibus ea, quæ ad hanc materiam spectant pluri-
mis & argumentis ac experimentis exactissime
absolutissimeq; tractantur et explicantur.*
Omnia nunc diligenter recognita & emen-
datius quam ante in lucem edita, aucta & figu-
ris illustrata operâ & studio
Wolfgangi Lochmans / I.U.D.
& Mathemati:
*Ad calcem libri adjunctus est Index Capi-
tum Rerum et Verborum locupletissimus*
EXCVSVS SEDINI
Typis Götzianis Sumptibus
Authoris
Anno M.DC.XXVIII.

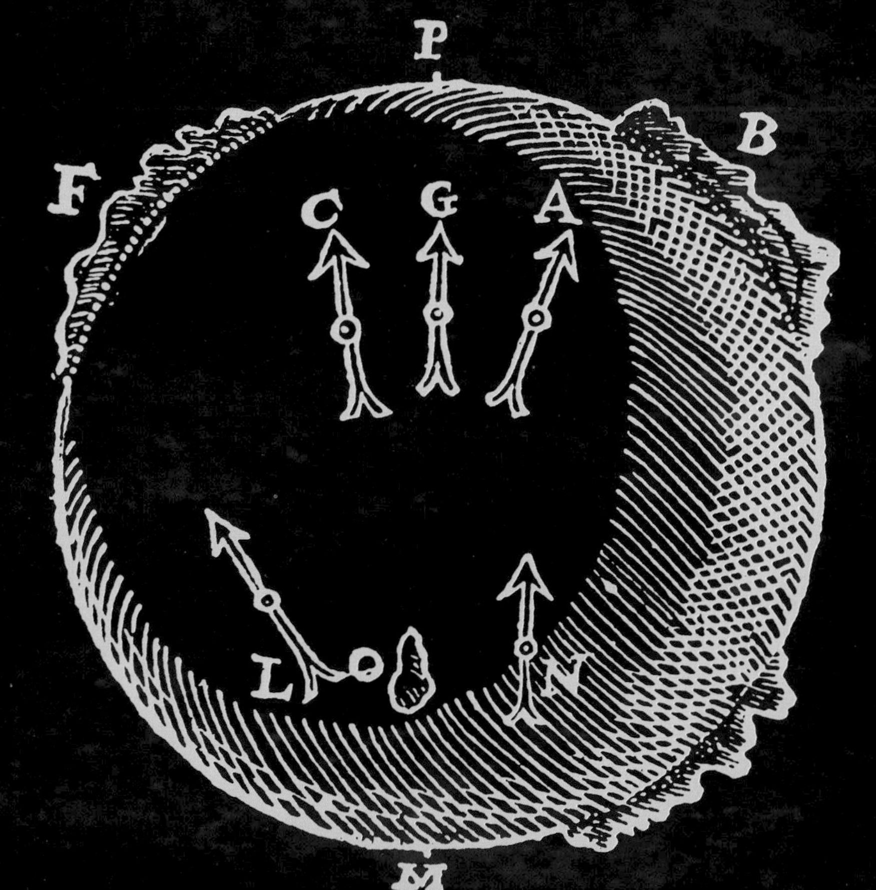

opposite poles of two magnets attract each other, so the points of a compass needle are pulled towards Earth's poles, showing the directions north and south. Gilbert proved this to be the case using a terrella, a small model world carved from a lodestone. A compass placed on the surface of the terrella behaved just as it would when used on Earth itself.

Gilbert was the first to suggest (correctly) that under the Earth's rocky surface the planet must contain a lot of magnetic iron. His work on the force of magnetism was the start of a drive to understand the forces that held the elements together.

21 Francis Bacon's New Method

AS NATURAL PHILOSOPHERS (NOT QUITE SCIENTISTS YET) BEGAN TO EMBRACE EVIDENCE AS THE WAY TO UNDERSTAND THE COSMOS, AN ENGLISH LAWYER SUGGESTED A NEW WAY OF WORKING. His system became the first scientific method.

Francis Bacon was more of an ideas man than a hands-on researcher. He had a checkered career as a barrister (trial attorney), politician, and courtier to Elizabeth I and James I, her successor in 1603. As an old man, he suffered a spectacular fall from grace, accused in 1621 of financial corruption, which saw him spend time in the Tower of London. He spent his last few years out of the public gaze, and after his death gossip spread that he had been the king's lover.

Today Bacon is better remembered for his invaluable contribution to science. In 1620 he published *Nova Organum Scientiarum* (A New Instrument of Science), in which he outlined a system of logic which he contended was more effective than that of Aristotle's. The Baconian method started with a reduction of the problem to describe it in the simplest terms possible. Next Bacon suggested doing away with syllogisms set out by Aristotle and other Greek philosophers. Their type of deductive logic arrives at truth by combining two premises. For example 1) All men die, 2) Francis Bacon was a man, therefore Francis Bacon died. Deduction works fine as long as the premises are correct, but one error leads to a cascade of further mistakes. Bacon suggested using inductive reasoning instead. In this case an explanation is proposed for an observed phenomenon. Unlike deduction, the proposition is not automatically regarded as correct, but must be shown to be true or false by a test, or experiment. Bacon's work had huge influence as the rallying call for the Scientific Revolution that was to come

TOP LAWYER

In Britain and Commonwealth countries (ex British Empire) the most senior lawyers are appointed Queen's Counsel—known as QCs. (KC is used when the monarch is male.) Francis Bacon became the first ever QC in 1597, largely as a concession for missing out on promotions to more powerful positions.

Francis Bacon's book Nova Organum *(A New Instrument) was titled in reference to* Organon *(The Instrument), Aristotle's book of logic.*

22 Robert Boyle: *The Sceptical Chymist*

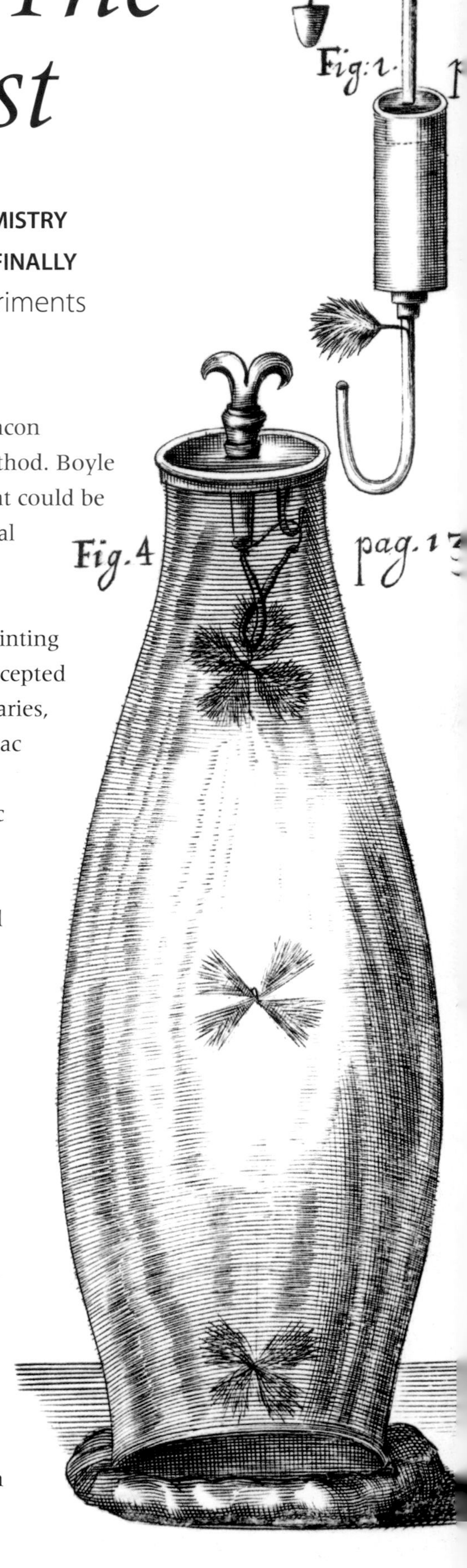

THE
SCEPTICAL CHYMIST:
OR
CHYMICO-PHYSICAL
Doubts & Paradoxes,
Touching the
SPAGYRIST'S PRINCIPLES
Commonly call'd
HYPOSTATICAL;
As they are wont to be Propos'd and
Defended by the Generality of
ALCHYMISTS.
Whereunto is præmis'd Part of another Discourse
relating to the same Subject.

BY
The Honourable ROBERT BOYLE, Esq;

LONDON,
Printed by J. Cadwell for J. Crooke, and are to be
Sold at the Ship in St. Paul's Church-Yard.
MDCLXI.

The title page of The Sceptical Chymist *by Robert Boyle.*

THANKS TO NEW SCIENTIFIC METHODS, CHEMISTRY STEPPED OUT OF ALCHEMY'S SHADOW AND FINALLY CAME OF AGE. One man's systematic experiments on the nature of air led the way.

Robert Boyle was born the same year Francis Bacon published his seminal work on the scientific method. Boyle grew up to be perhaps the best exponent of what could be achieved using a systematic approach to chemical research. It was Boyle's own book, *The Sceptical Chymist*, published in 1661, that banished the hocus-pocus and superstitions of alchemy by pointing out the contradictions and errors that littered accepted thinking. The work of Boyle and his contemporaries, such as Denis Papin, Robert Hooke and even Isaac Newton, did much to replace alchemy with a new field called chemistry, a rigorously scientific investigation of nature's substances.

Although Boyle's views joined the chorus of voices that were advocating that nature was comprised of more than the four classical elements, Boyle still believed in the transmutation of matter and never stopped searching for a way to multiply gold from iron. However, he disputed that transmutation was influenced by the paranormal, and argued strongly that like other phenomena, it was best investigated scientifically. (Of course 250 years later, transmutation was indeed discovered to be possible through radioactivity, although not in a way that Boyle might have hoped.)

Air-pump experiments

Robert Boyle was the son of an Anglo-Irish earl, which afforded him the best education as a boy and young man. However, the English Civil War in the 1640s made life more difficult for his Royalist family. Nevertheless he remained comfortably off, if not fabulously wealthy, and was able to set up a laboratory in London.

Boyle employed Robert Hooke as his assistant, tasking him to build a version of a pump recently invented by the German Otto von

Diagrams of Boyle's experiments with air and vacuums, showing the sealed glass vessels built especially for the purpose. On the left, the diagram shows feathers falling in a vacuum.

Guericke, which could pull air out of any vessel, creating a vacuum.

(Young Hooke was destined for a long career in science, including discovering and naming biological cells and is also known for Hooke's law of elasticity.)

Like most 17th-century scientists, Boyle regarded air to be a single substance. (It is of course a mixture of gases.) His early experiments showed that sound could not travel without air, flames would not burn without it, and animals and plants could not live in the absence of air.

Robert Boyle (right) and his French assistant Denis Papin. One of their famous spherical air pumps is located behind the pair.

The pump experiments also resulted in the discovery for which Boyle is most remembered: Boyle's law. This states that the pressure of a gas is inversely proportional to its volume. Compressing a quantity of gas into a smaller volume results in its pressure increasing. This is perhaps a rather intuitive fact, but as one of the laws described the behaviors of gas, it forms a foundation stone of atomic theory.

The nature of air

Robert Boyle promoted the idea that air was made up of corpuscles, tiny units that moved in all directions, bouncing off each other and spreading out until they collided with the walls of their container. Any substance that behaved in this way was known as air. (The word *gas* was not yet in common usage.) Nevertheless Boyle noticed different characteristics. For example the "air" that bubbled off from metals placed in mineral acids caught fire when a candle was placed in it. Obviously candles did not burn air as a general rule, and the difference was put down to the purity of the samples. Room air was assumed to be contaminanated, while the air released from metal could only be pure. (It was actually hydrogen.) At the time bad air was seen as the cause of diseases, and Boyle was a sickly person. However, his habit of tasting every substance that he investigated is a more convincing reason for his frequent ill health.

THE INVISIBLE COLLEGE

Robert Boyle was a chief figure in a group of scientists who gathered to discuss their work and to help each other solve problems. But, while later scientific societies were set up to disseminate knowledge, Boyle's group appears to have been rather secretive. The habits of alchemy die hard.

23 Phosphorus, the Light-Giver

IN THE 17TH CENTURY, THE CONCEPT OF AN ELEMENT HAD BEEN LARGELY UNCHANGED SINCE THE GREEKS. Then one of the last alchemists made a rather startling discovery.

Modern chemistry has revealed that there are around 90 naturally occurring elements on Earth. For many of the most familiar, such as gold, copper, or sulfur, we have no named discoverer. That changed with Hennig Brand, a glass maker, merchant, and alchemist from Hamburg in northern Germany. In 1669 he discovered phosphorus, the first named person in history to have discovered a new element.

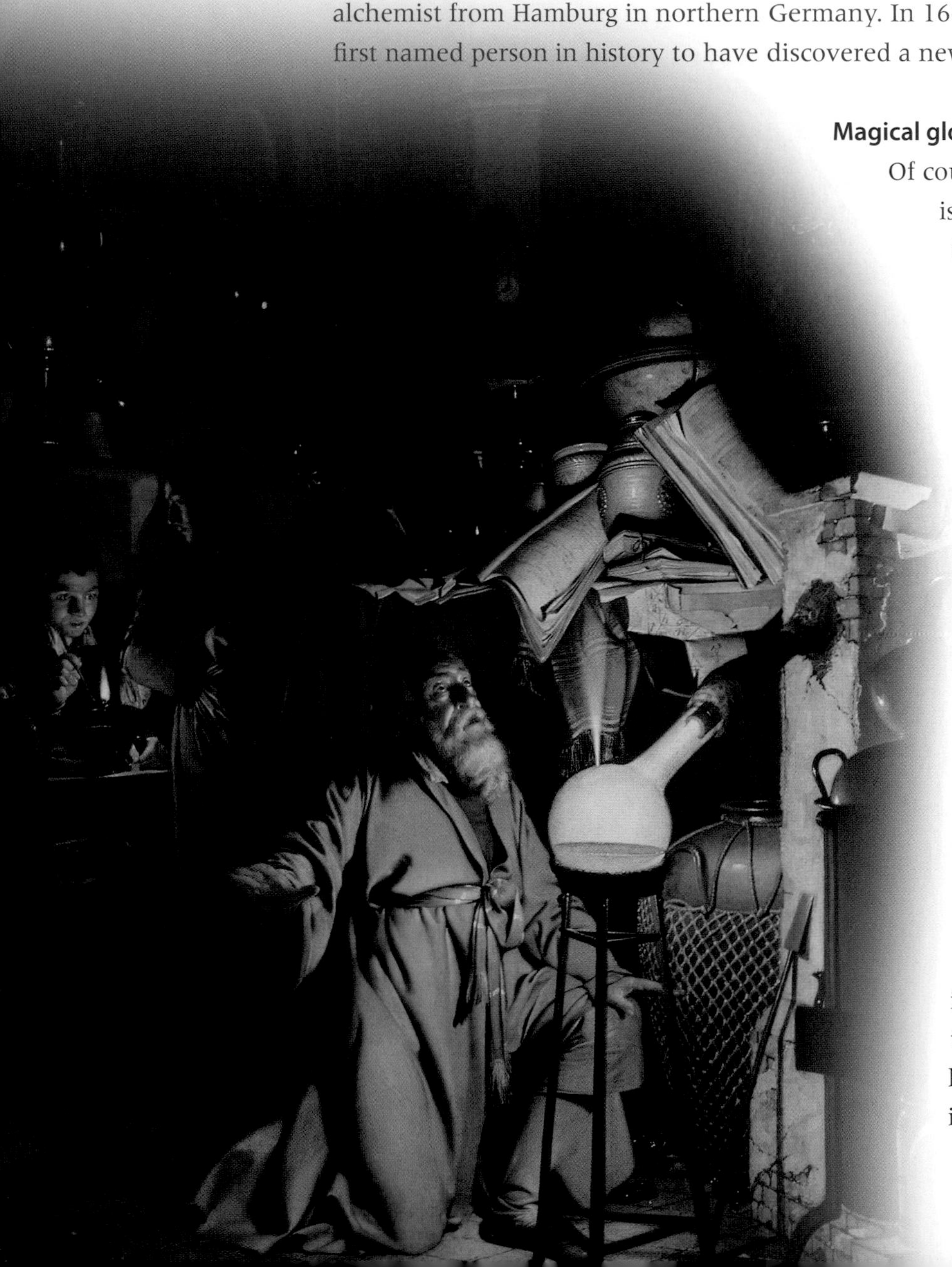

Magical glow

Of course, Brand had no idea that is what he had done. Earlier figures, such as Paracelsus in the 1500s, had pushed the idea that sulfur, mercury, and salt were primal substances that had an influential role in the formation of materials, without actually being elements like earth, fire, water, and air. Robert Boyle dismissed Paracelsus' convoluted theory but agreed that there must be more than four elements—science would reveal them in time.

Little is known about the life and career of Hennig Brand, so we are unsure if he was conversant with the cutting-edge arguments of the time. Whatever his scholarship, it is hard to imagine what goes through a man's mind after he has boiled a vat of urine for hours, if not days, only to find that the

The glow produced in Brand's flask was caused when the phosphorus reacted with oxygen in the air. Robert Boyle found that sealing the phosphorus in a flask caused the glowing to fade (as the oxygen ran out).

resulting deposit then begins to glow in the dark. He must have thought that his search for the philosopher's stone had come to an end with this magical white luminous substance, which he named *phosphorus mirabilis*.

Elemental recipe

Brand published his recipe for his miraculous discovery. He is thought to have used more than 1,000 liters (26 gallons) of urine to produce less than 100 grams (3.5 ounces) of phosphorus. First he let the urine fester until it smelt suitably nasty. Then he boiled it down to a paste, distilled off a red oil, and reduced the rest into a black porous material and a white salt. He discarded the salt, recombined the oil and spongy material and heated them for 16 hours. He passed the fumes through water, perhaps hoping to see gold. Instead he got phosphorus. (Urine is naturally high in phosphates, compounds of phosphorus and oxygen.) Modern analysis shows that if he had used the white salt he would have yielded much more phosphorus—and there was no need to leave the urine to go bad; fresh ingredients were fine.

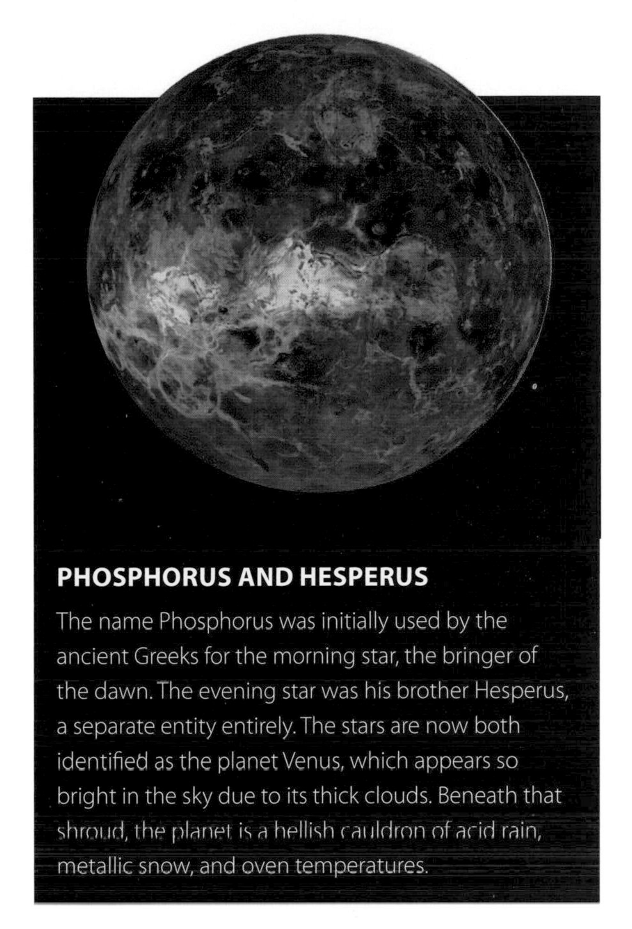

PHOSPHORUS AND HESPERUS

The name Phosphorus was initially used by the ancient Greeks for the morning star, the bringer of the dawn. The evening star was his brother Hesperus, a separate entity entirely. The stars are now both identified as the planet Venus, which appears so bright in the sky due to its thick clouds. Beneath that shroud, the planet is a hellish cauldron of acid rain, metallic snow, and oven temperatures.

24 Multiplying Metals

WHEN ROBERT BOYLE ENCOURAGED THE BRITISH PARLIAMENT TO REPEAL THE BAN ON ALCHEMY IN 1689, ISAAC NEWTON WAS SUSPICIOUS. He felt sure that his old friend was on the verge of multiplying gold.

The Mines Royal Act of 1689 was passed the year after the Glorious Revolution, a nearly bloodless coup that saw the British throne taken over by Mary, daughter of James II, and her husband William of Orange, a Dutch nobleman. As well as removing the controls on alchemy, which so intrigued Newton, the act ended the royal monopoly on mines, making it possible for anyone to extract base metals without interference. (However, if anyone struck gold or silver, the King's men would come for it.) As a result metal industries boomed, especially those of iron and brass (an alloy of copper and zinc), driving technological development that put the Industrial Revolution only decades away.

In the end Newton's jealousy of Boyle was misplaced (not for the first time). However, according to philosopher John Locke, Boyle did give Newton a mysterious red earth before he died, saying that it turned mercury into gold. Whether it did was never recorded.

25 The Great Fire

IN 1660, THE ROYAL SOCIETY OF LONDON BECAME THE WORLD'S FIRST OFFICIAL ORGANIZATION DEVOTED TO SCIENCE. Ironically, when London burned down six years later, narrowly missing the society's meeting rooms, none of its members could explain what fire actually was.

In classical terms, fire was viewed as a substance, only seen when it was released from mixtures. Paracelsus developed the idea further, saying the reason why some things burned—such as the wooden houses of 1660s' London—was because they contained sulfur. Air was not involved in the process other than as the medium through which the heat and flames could pass. Water on the other hand mostly blocked fire.

Finding flaws

This theory had long been questioned by metalworkers. If fire was something being released, why did metal get heavier when it was heated to very high temperatures? (We now understand that oxygen reacts with the metal, adding to its total weight.)

An answer was offered by the German physician John Joachim Bercher. The year after the Fire of London, he suggested that a substance called phlogiston (coined from the Greek word for burn) was being released during combustion. When challenged over the weight gain seen in metal, he offered a mind-boggling solution: Phlogiston weighed less than nothing and so when it was released from objects they got heavier!

The Great Fire of London in 1666 was started in a bakery. Robert Hooke, the secretary of the Royal Society, built a monument at the source of the fire in Pudding Lane. The monument was a pillar that also served as a telescope.

26 Judging Temperature

SCIENCE NEEDS PRECISION AND THAT MEANT MEASURING THE FIRE, PHLOGISTON, OR WHATEVER IT WAS THAT PRODUCED HEAT. Thermometer technology was developing fast by the 1700s; all that was needed were some units.

No one knows who invented the thermometer. It uses the principle that fluids expand as they get warmer and contract when they are cold. This fact was known by Hero of Alexandria in the 1st century AD, but the first records of working thermometers (using water) appear in the early 17th century. The scale of a thermometer used to measure and compare temperatures is entirely arbitrary, but it took until 1724 for a practical scale to be developed by Daniel Gabriel Fahrenheit.

Fahrenheit was a glassblower who invented alcohol and mercury thermometers. Temperature scales are formulated between an upper and a lower point, both of which are constant and easily obtainable so new thermometers can be calibrated. They also need to be close to standard conditions so they are a practical measure in everyday situations. Fahrenheit chose human body temperature as the upper point, set at 96°, and a mixture of ice, water, and ammonium chloride, which always stabilizes to the same temperature was the lower point of 0°. In 1742, Anders Celsius developed a simpler scale based on the melting and freezing point of water. This scale is still used in most parts of the world and the Celsius degree in the unit of choice for all scientists.

MERCURY THERMOMETER

Fahrenheit's scale was a close copy of one proposed by Newton. However, its success was due to Fahrenheit's use of mercury in his thermometers. Mercury expands in small amounts so a working device need not be huge. Mercury's rate of expansion is also very uniform, with every increment proportional to the increase in temperature. (The rate of expansion of other liquids varies at different temperatures.) Mercury thermometers remained the most accurate until the invention of digital devices.

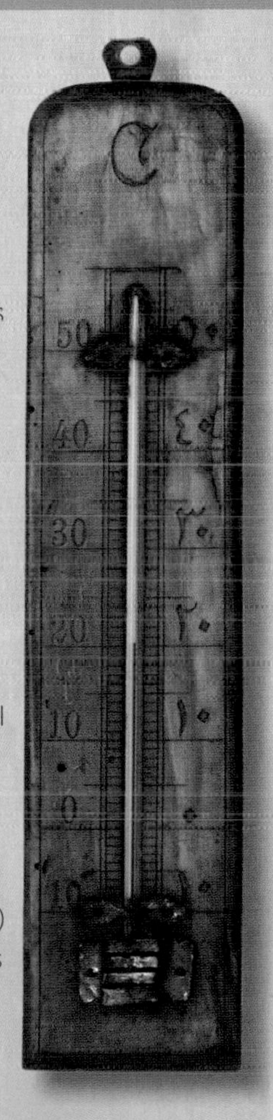

27 Storing Electricity

BY THE 18TH CENTURY, RESEARCH INTO ELECTRICITY HAD HIT A ROADBLOCK. Primitive generators could produce sparks, but there was no way of storing the mysterious "electrical fluids."

One of Otto von Guericke's inventions, the vacuum pump, had been behind Robert Boyle's discoveries on air, and a second now played a role in the field of electromagnetism. In 1660 von Guericke made an electrostatic generator from a carved ball of sulfur that rotated on a wooden handle. Spinning the ball by hand, like rubbing a piece of amber, made it attract objects and release small sparks.

This and other "friction machines"—later models used glass instead of sulfur—were largely toys. A common party trick in the 1740s was the "electric kiss," where a person stood on a stool (insulating them from the ground) and was charged by the generator. A light kiss from another participant was enough to discharge the static, literally creating a spark between the couple.

In 1745 Ewald Georg von Kleist lined a jar inside and out with silver foil, keeping the two layers apart. He connected the inner one to a generator, while the outer layer was grounded. Electricity built up inside and could be released by touching the two foils together. At first Kleist did this by hand—and was lucky not to die in the process. A similar device built by Pieter van Musschenbroek was shown at the University of Leiden in the Netherlands the following year and so it became known as a "Leyden jar." Leyden jars remained the main source of electricity for around 60 years.

In the 18th century electricity was thought to be a fluid, so it seemed logical to bottle it. The charge was stored on the large internal surface of the jar, working like a capacitor does in modern electronics.

ONE FLUID THEORY

It had been noted that some charged objects repelled each other, for example amber pushed away glass. It had been thought that these materials contained two types of electrical fluid. However, Benjamin Franklin suggested one had lost fluid while the other had gained it, and thus introduced the concept of positive and negative charge. We also have Franklin to thank for the word *battery*, which he used to describe a set of Leyden jars—comparing them to group of cannons.

Benjamin Franklin was lucky to survive his 1752 experiment where he charged a Leyden jar using the electricity in lightning.

28 Fixed Air

WHILE LOOKING FOR A CURE FOR KIDNEY STONES, A SCOTTISH DOCTOR STUMBLED UPON A NEW FORM OF AIR. He named the gas fixed air, but we know it better as carbon dioxide.

Joseph Black was researching a thesis for his medical training when he made his discovery of carbon dioxide in the 1750s. He had an interest in chemistry and so began a search for a mineral solvent that could dissolve kidney and gall stones inside the body, thus saving the pain of passing them. He knew a solution of quicklime would do it, but swallowing that would do more harm than good.

Releasing fixed air

So he turned his attention to the less potent magnesia alba, described as a mild akali at the time (and now known as magnesium carbonate). It had no impact on kidney stones (although Black noted its laxative and antacid properties in his records), but nevertheless the young Scot became intrigued with the bubbles of "air" that the white powder gave off when treated with acid. (Acid and carbonate reactions always produce carbon dioxide.) He heated the magnesia alba and found that although the resulting crystals looked the same, they no longer bubbled in acid. He surmised that the air "fixed" inside was released by the heat. (The carbonate had broken up into an oxide and carbon dioxide gas.)

Black was unable to collect any air released through heating, so instead he weighed the magnesia powder and the acid before mixing and compared that with their combined weight after the "fixed air" had been released. The process showed that weight had been lost.

SPIRIT OF THE WOODLAND

The first record we have of carbon dioxide comes from the Belgian Johannes van Helmont (below right, collecting supplies from an alchemist). He noticed that the ash left when charcoal had burned weighed much less than the original fuel. He collected the "air" that was released, calling it *spiritis sylvestris* (woodland spirit). Van Helmont also coined the word *gas*, probably deriving it from *chaos*.

A common substance

Black knew other mild alkalis such as limestone (calcium carbonate) also gave off a fixed air. He showed that these were in fact the same substance—both turned lime water, a solution of calcium hydroxide, cloudy. (The carbon dioxide was forming tiny particles of solid calcium carbonate.) The air released by burning, exhaling, and fermentation also did this, making Black the first person to prove that one substance could be produced from several sources.

29 Revealing Latent Heat

Joseph Black attended Edinburgh University as an undergraduate and was a professor of medicine at Glasgow University. Both Scottish institutions have named their chemistry facilities in his honor.

JOSEPH BLACK'S WORK ON FIXED AIR IS NOW OVERSHADOWED by his research into heat, which formed the basis of a science called thermodynamics.

Later in his career, while carrying on his day job as a physician, Joseph Black became interested in the effects of heat (and cold) on substances. He took as his subject the way water changed into ice and steam.

Hidden heat

It was well understood that adding heat made ice melt and water boil, but Black noticed that heating ice as it is melting does not raise its temperature but results in more water. Similarly the temperature of boiling water does not increase, even when heated continually. Instead the water becomes steam of the same temperature.

Black concluded that the heat energy was being used to transform the solid into a liquid and liquid into gas, rather than merely making it warmer. Once the sample has changed state completely, for example all the ice had melted, the addition of heat goes back to boosting the temperature. He called this phenomenon latent heat. The latent heat of fusion is needed to melt a substance, while the latent heat of vaporization is concerned with evaporation. Although Black did not understand it in these terms, the process is identical in reverse. Ice melts when latent heat is added to it, and when it freezes it loses the same amount of latent heat. Therefore at the freezing point, the water gives out heat as it rearranges into ice, which keeps the temperature constant.

Heat capacity

Black also noticed that some substances, largely liquids, warmed up more slowly than others, despite being heated in the same way. He named this phenomenon heat capacity and noted that water had a high heat capacity compared with other liquids such as alcohol. Fellow Scot James Watt was a friend of Black, and used this notion of heat capacity to greatly improve the efficiency of steam engines. He did this largely by separating the boiling and condensing process. In early designs, time and heat were wasted in frequently reheating the water.

WHY ICEBERGS FLOAT

Because water is so common it is hard to imagine that it is very unusual stuff. One of its strange characteristics is that ice is less dense than water. Just about everything else gets denser when it freezes. As water turns to ice, the molecules spread out into a orderly pattern so they can bond together. On a large scale that means ice floats, and water freezes from the top down. If ice sank, most of the seabed would be permanently frozen, changing the nature of Earth considerably.

30 Hydrogen: Air that Burned

HYDROGEN IS THE MOST COMMON ELEMENT IN THE UNIVERSE, BUT IT REMAINED UNKNOWN UNTIL 1766. Then Henry Cavendish found a flammable air that bubbled up when acid reacted with iron.

Henry Cavendish was scientific nobility, the son of a lord who was also a leading scientist in his own right. Young Henry was able to build a laboratory in his father's house. His apparatus for collecting hydrogen involved heating metal filings in acid and running the gas produced into an upturned vessel placed in water.

One of the first experiments performed in high-school chemistry class is seeing what happens when a piece of metal is added to an acid. The result is a satisfying fizz of gas, a small sample of which will burn with a characteristic squeak—hydrogen. It suffices to say nowadays that acids and metals react to produce hydrogen but to the first chemists this was not so straightforward. Of the metals they had close to hand—copper, silver, and iron—only the latter really reacts like this. (Chemistry students today usually use magnesium, which was as yet undiscovered.)

In the 1660s Robert Boyle had found that the air produced by iron filings in acid burned easily, but it was another century before Henry Cavendish realised that this air was a distinct substance. He called it "flammable air," and suggested this lightweight substance was the "phlogiston" that was thought at the time to be the cause of fire itself.

LIGHTER THAN AIR

Hydrogen is the lightest gas in the Universe, weighing 16 times less than oxygen, for example. Hydrogen balloons were pioneered less than 20 years after Cavendish isolated the gas. However, the promise of this new form of transport was lost when the Hindenburg airship, the largest flying craft of all time, burst into a ball of flames in 1937 in New Jersey. As well as being the lightest, hydrogen is also the most flammable gas.

31 Phlogisticated Air

ACCORDING TO THE PHLOGISTON THEORY OF COMBUSTION, AIR THAT PROMOTED BURNING DID SO BY ABSORBING a mysterious material called phlogiston from a flaming substance. While investigating the role of Black's new "fixed air" in combustion, Daniel Rutherford ended up discovering a new gas altogether.

It was a commonly held notion in the 18th century that life was sustained by breathing in "good" air, which then was somehow converted into "bad" air, which was breathed out. Rutherford correctly reasoned that Joseph Black's "fixed air" (carbon dioxide) was the bad component. Indeed Black had shown that living things expired in fixed air and flames went out. Rutherford's 1772 investigation began with preparing a sample of fixed air by keeping a mouse in a jar. He then burned a candle in the sample to remove any good air that remained. Next he removed the fixed air, in the way Black had shown by bubbling it through lime water. This, he theorized, would turn the bad air "good" again, but he found that flames still died in the remaining gas. Turning to the phlogiston theory for an explanation, Rutherford termed his sample as "phlogisticated air," meaning it was saturated with phlogiston and could not support further burning. In fact, he had discovered the main ingredient of air, an inactive gas now called nitrogen.

32 Joseph Priestley's Airs and Graces

Joseph Priestley's work had a lasting impact. It was Priestley who noticed that a solid polymer made from the milky latex from certain plants could be used to rub out pencil marks. The substance has been called rubber *ever since.*

THE SCIENTISTS WHO STUDIED GASES, OR AIRS, BECAME KNOWN AS PNEUMATIC CHEMISTS, AND JOSEPH PRIESTLY WAS THE NEXT ONE TO MAKE HIS MARK. Not only did this part-clergyman, part-chemist discover what is now known as oxygen, his political and religious views resulted in him being hounded out of his native country.

Joseph Priestley's first calling was as a Presbyterian minister. Although not a great public speaker, Priestley gave his fair share of sermons, using them to argue in favor of independence for the American colonies, which were on the verge of revolution. The outspoken Priestley was not a popular churchman and frequently moved in search of new congregations, all the while developing his hobby of experimental chemistry.

An engraving from Priestley's book Experiments and Observations on Different Kinds of Air *shows the apparatus he used to study gases.*

Building on bubbles

When Priestley moved to Leeds in northern England in 1770, he lived next door to a brewery. The fermentation process central to brewing produced the "fixed air" recently described by Joseph Black. Priestley found that this gas, when dissolved in water, made a refreshing sparkling drink. The invention of soda water elevated Priestley to the highest scientific circles, and he was recruited by the Earl of Shelburne as a science adviser and secretary.

The new position afforded him the time to perform more pneumatic chemistry. His first experiments were into the nitrous airs, gases given off when nitric acid reacts with certain metals. One of the colorless airs Priestley isolated he named it dephlogisticated nitrous air—seemed to remove the "goodness" from regular air, making it unable to support fire. (This gas was nitrous oxide, or laughing gas, which was actually reacting with the oxygen in the air.)

In 1791, rioters burned down Priestley's house in Birmingham, England. While living in Birmingham, Priestley attended the Lunar Society, a social group for industrialists and scientists, which included Josiah Wedgewood, James Watt, and Erasmus Darwin.

Further rise and then fall

In 1774, Priestley was able to collect another air by heating mercuric oxide. This gas made candles burn more brightly and smoldering charcoals burst into flame. He named it dephlogisticated air since it promoted combustion—or, according to the theory of the day, absorbed phlogiston from burning materials. Priestley reported his work to Antoine Lavoisier, the great French chemist who would soon rename the new gas oxygen.

After leaving Shelburne's employ, Priestley soon found himself back in hot water. With America lost and France in revolution, he became the target of an English nationalist backlash. He opted to move to Pennsylvania in 1794, where he discovered a fourth and final gas—carbon monoxide—which burned with a characteristic deep blue flame.

SODA WATER

One person who saw the commercial potential of Priestley's soda water was German jeweler Johann Jacob Schweppe. He opened the first sparkling mineral water factory in Geneva in 1783. His name has been associated with soda ever since.

33 Scheele: Unknown Discoverer

AWAY FROM THE SCIENTIFIC SALONS OF 1770S PARIS AND LONDON, CARL SCHEELE, A SWEDISH PHARMACIST was making many discoveries that advanced chemistry, some of which ended up being credited to others.

Despite being ignored by his wealthier scientific rivals, Scheele was not forgotten. The principle ore for tungsten, discovered in 1821 in Sweden, was named scheelite in his honor.

Although Priestley and Lavoisier loudly disputed it at the time, it is likely that Carl Scheele isolated and investigated oxygen in 1772—he just did not tell anyone. By the time he finally made his work public, the qualities of *fire-air*, as he called it, were already well established by others.

Finding fire-air

Like Priestley, Scheele arrived at oxygen by studying the nitrous airs, the gases given off by nitric acid. His ideas were based on the phlogiston theory. He thought that a component of air (the so-called fire-air) was combining with the phlogiston in materials to make them burn and release heat. He noticed that both nitric acid and heat had the same effect on metals, turning them into the same earths (oxides in modern terminology). Scheele's hypothesis was that heat was the result of fire-air combining with the phlogiston in a metal. Since heat and the acid both did the same thing, perhaps nitric acid could reverse the process and take the phlogiston out of the heat, leaving the fire-air behind.

In what seems a lucky accident as much as scientific foresight, Scheele "fixed" the qualities of nitric acid in a solid (potassium nitrate), which he then heated, collecting any resulting gases. The nitrous airs were absorbed by calcium hydroxide chemicals, leaving the first sample of pure fire-air, or oxygen.

THE STORY OF TUNGSTEN

The name tungsten is a composite of the Swedish words *tung sten*, meaning "heavy stone." Nevertheless its chemical symbol is W because in German it is called *wolfram*, which means "wolf cream," the name asserted (for unclear reasons) by the 16th-century metallurgist Georgius Agricola. Tungsten has the highest melting point of any metal. In theory it would not melt inside a sunspot. A tungsten filament, glowing white hot, is what provides the light in an incandescent bulb.

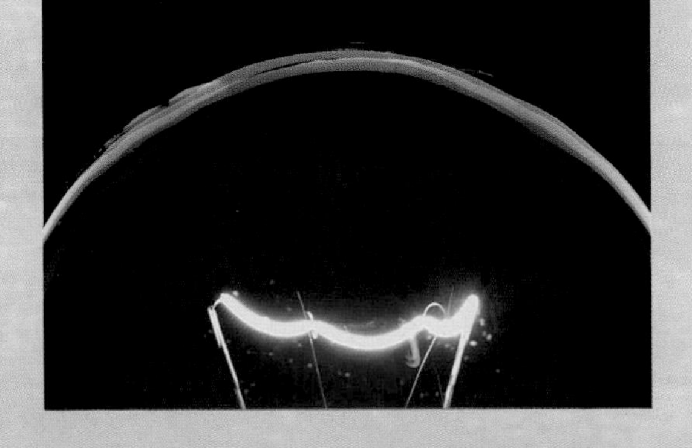

Other work

As well as his work in pneumatic chemistry, Scheele was also one of the most prolific discoverers of elements, although he would not have seen them as such. The Swede is credited with finding the metals barium (1774), molybdenum (1778), and tungsten (1781) in rock-forming minerals. Scheele also prepared chlorine gas in 1774, although this discovery is often credited to Humphry Davy who proved the gas was an element 37 years later.

34 Lavoisier's Simple Substances

ANTOINE LAVOISIER HOLDS THE TITLE OF FATHER OF CHEMISTRY, rightly so because his contributions to the science were as epoch-changing as the French Revolution that framed his life.

	Noms nouveaux.	*Noms anciens correſpondans.*
Subſtances ſimples qui appartiennent aux trois règnes & qu'on peut regarder comme les élémens des corps.	Lumière	Lumière.
	Calorique	Chaleur. Principe de la chaleur. Fluide igné. Feu. Matière du feu & de la chaleur.
	Oxygène	Air déphlogiſtiqué. Air empiréal. Air vital. Baſe de l'air vital.
	Azote	Gaz phlogiſtiqué. Mofete. Baſe de la mofete.
	Hydrogène	Gaz inflammable. Baſe du gaz inflammable.
Subſtances ſimples non métalliques oxidables & acidifiables.	Soufre	Soufre.
	Phoſphore	Phoſphore.
	Carbone	Charbon pur.
	Radical muriatique.	Inconnu.
	Radical fluorique	Inconnu.
	Radical boracique	Inconnu.
Subſtances ſimples métalliques oxidables & acidifiables.	Antimoine	Antimoine.
	Argent	Argent.
	Arſenic	Arſenic.
	Biſmuth	Biſmuth.
	Cobolt	Cobolt.
	Cuivre	Cuivre.
	Etain	Etain.
	Fer	Fer.
	Manganèſe	Manganèſe.
	Mercure	Mercure.
	Molybdène	Molybdène.
	Nickel	Nickel.
	Or	Or.
	Platine	Platine.
	Plomb	Plomb.
	Tungſtène	Tungſtene.
	Zinc	Zinc.
Subſtances ſimples ſalifiables terreuſes.	Chaux	Terre calcaire, chaux.
	Magnéſie	Magnéſie, baſe du ſel d'Epſom.
	Baryte	Barote, terre peſante.
	Alumine	Argile, terre de l'alun, baſe de l'alun.
	Silice	Terre ſiliceuſe, terre vitrifiable.

Lavoisier's Table of Simple Substances in its final 1789 form contained 33 items, 25 of which are now recognized as elements.

It is Lavoisier whom we have to thank for the very concept of a table of elements, but that was one achievement among many. Two advantages, combined with his undoubted genius, helped Lavoisier take the chemistry crown. Firstly, he used his great wealth to have exquisite apparatus crafted with the utmost precision, and secondly he stole the ideas of others, passing them off as his own.

In the fall of 1774 he had a dinner-table chat with Joseph Priestley in Paris about dephlogisticated air, and almost simultaneously received a letter from Carl Sheele about the discovery of fire-air. Nevertheless, in 1777, Lavoisier pronounced to the world a new gas called *oxygène*. His name meant "acid former," from *oxux* meaning sharp. Lavoisier's contention that this gas was present in all acids was false, however.

Oxygen was added to Lavoisier's List of Simple Substances, materials he believed could not be further divided, a precursor to the periodic table of elements. There were several errors—the gas section of the list had entries for light and *caloric*, an essential form of heat and fire—but the modern idea of chemical elements had been born.

Lavoisier operates his solar furnace, an immense lens that focused sunlight to combust samples without fear of contamination from fuels.

35 The Conservation of Matter

One of Lavoisier's most celebrated moments was transmuting air into water. True to form, he was not the first to do so but he used the feat to highlight a fundamental law of chemistry.

Both Henry Cavendish and Joseph Priestley had noted that "flammable air" left droplets of water on the inside of glass vessels after it had burned. As a result Lavoisier renamed the gas *hydrogen*, meaning water-former. Cavendish noted that the water that formed tasted slightly sharp (acidic), and Lavoisier combined this fact with his theory that the presence of oxygen gas was the cause of acidity. (Actually the sharp flavor was caused by a trace of nitric acid that was formed by the intense heat of the combustion from the otherwise inert nitrogen in the air.)

Lavoisier's chemical scales were more precise than those of his contemporaries, so he could confidently confirm that the total weight of matter was unchanged during reactions.

Disproving Aristotle

Despite the flaw in his acid hypothesis, Lavoisier set fire to a mixture of pure hydrogen and oxygen. Instead of acid he found only water. He realized immediately that here was proof that would consign Aristotle's theory of elements to history: water was not an element at all but a compound of two gases. Since his explanation only referred to hydrogen and oxygen, Lavoisier's discovery also ended the concept of phlogiston being central to combustion.

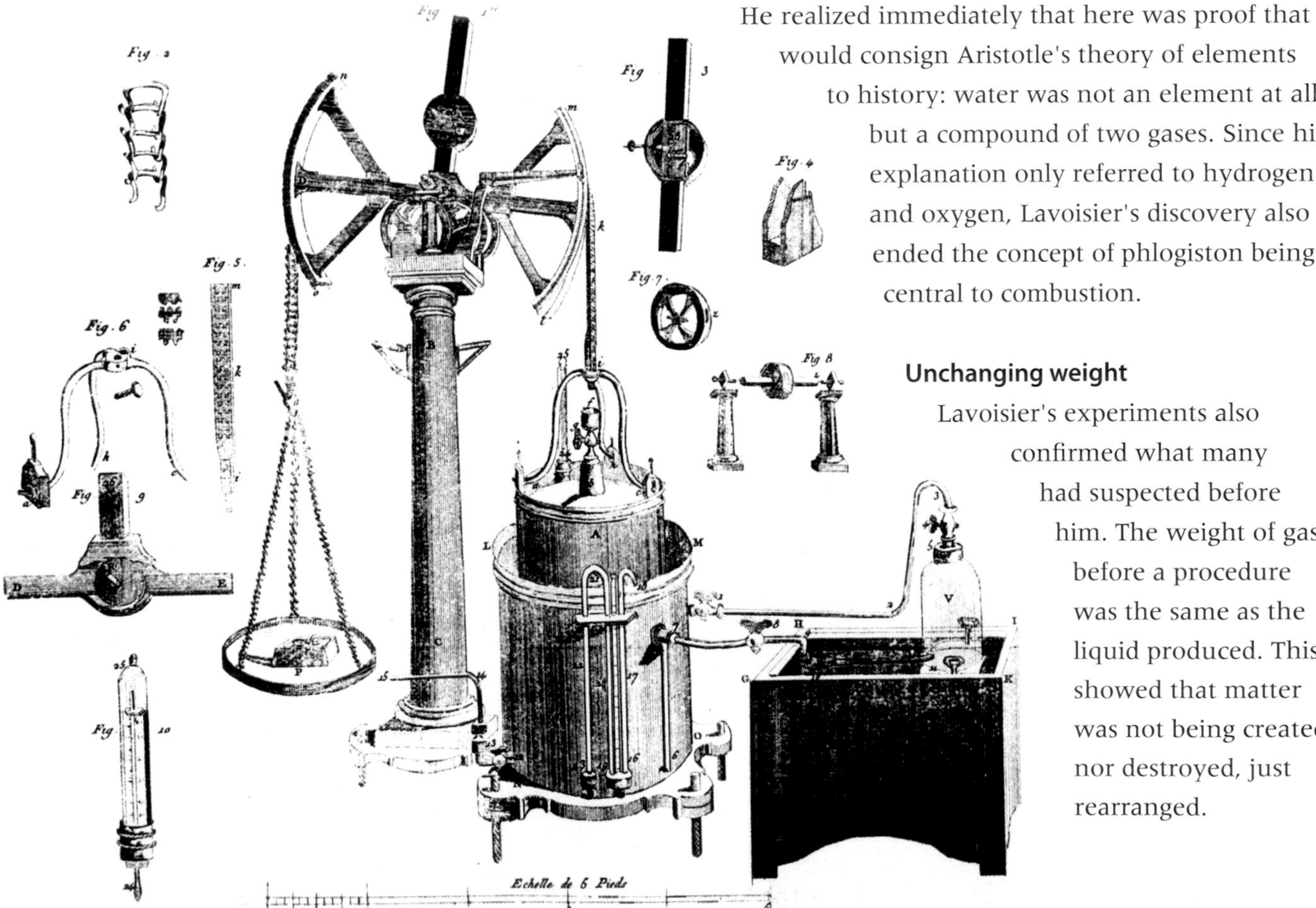

Unchanging weight

Lavoisier's experiments also confirmed what many had suspected before him. The weight of gas before a procedure was the same as the liquid produced. This showed that matter was not being created nor destroyed, just rearranged.

METRIC SYSTEM

Lavoisier's fame reached its peak just as the French Revolution began to rage through the streets of Paris. Following the removal of the king, Lavoisier's skills were not wasted by the new republic. He was appointed to the Commission for the Establishment of the Metric System, with a remit to make units of weight and distance uniform across France—and eventually the world.

36 Measuring Heat

ANTOINE LAVOISIER WAS NOT THE ONLY ARISTOCRATIC GENIUS IN 18TH-CENTURY FRANCE. He teamed up with mathematician Pierre-Simon de Laplace to investigate heat and light.

Having disproved the phlogiston theory of combustion, which stated that fire was the release of a material called phlogiston, Lavoisier quite rightly wondered again about the nature of the heat and light contained within flames.

FOOD ENERGY

The energy in food is quantified by measuring the heat it releases when it is burned in a modern calorimeter. The unit used, the *calorie*, is named for Lavoisier's "fire matter."

Laplace is better remembered as a pioneer of statistical theory and probability.

Theory of substance

Light had long been a subject of investigation, and a number of theories abounded. The century before, Christiaan Huygens had promoted the theory that light was a wave, based on a demonstrable similarity between its behavior and that of other waves. Newton's interest in optics a few decades later resulted in the then novel concept of a *spectrum* of light made up of seven colors. (Six would have been easier, but the surprisingly mystical Newton thought seven was a more auspicious number, so he invented the color indigo.) Newton threw his weight behind a corpuscular theory of light, saying it was made of tiny weightless particles. Lavoisier took a similar approach to heat. He named the weightless material in fire as *caloric*, and unlike phlogiston, he could measure it.

Calorimeter

In the 1780s Lavoisier enrolled the services of Laplace to develop a caloric measurer, or calorimeter. The device had a central chamber that was surrounded by a packed layer of ice crystals. The *caloric* released by samples in the center melted a portion of the ice. The quantity of water produced was a measure of caloric. Laplace and Lavoisier measured the caloric released by burning charcoal. They then placed a guinea pig into their machine to compare the heat it released as it breathed. Their two results tallied, confirming that animal respiration was a type of combustion.

The calorimeter was a set of concentric vessels designed to isolate it from outside influence.

37 Coulomb's Law

WHILE LAVOISIER WAS MEASURING THE ENERGY CONTAINED IN FIRE, another French scientist was developing a means to quantify the strength of electrostatic forces.

By the 1780s it was agreed that electrical forces were the result of an imbalance in a fluid-like substance present in all things. Electrical behavior arose when there was a an excess or dearth of the fluid. Objects in opposite states (containing too much or too little) attracted each other, while those with similar quantities of electrical fluid exerted a repulsive force.

The electrostatic forces that were possible at the time were only weak, and in 1784 Charles-Augustin de Coulomb devised the torsion balance, which could measure forces of this magnitude. It involved a charged metal bar hanging from a thread. Electrical force from another charged object made the bar swivel around the thread. Coulomb found that the size of the movement was inversely proportional to the square of the distance between the objects. This relationship, now known as Coulomb's law, was the first hint of the workings of electrical force fields.

A larger version of Coulomb's torsion balance was used by Henry Cavendish to measure gravitational forces in 1798.

38 Analyzing Nature

WHILE THE NEW BREED OF SCIENTIST WAS REVEALING THE INNER WORKINGS OF NATURE, other researchers were focused on finding out what materials were out there in the rocks and clays that make up our world.

One of the main functions of chemistry to this day is to analyze unknown substances. At the end of the 18th century many of the minerals in rocks were just that, unknown substances. A German pharmacist (later professional chemist) called Martin Klaproth began applying the discoveries of Lavoisier, Cavendish, and others to find out what the world was made of.

In 1791, Klaproth found a new metal in a mineral called rutile. He named it titanium after the Titan gods of Greece. (Nevertheless, English mineralogist William Gregor is credited with identifying the metal first.) There is no doubt, however, that Klaproth was the discoverer of uranium (named for the planet Uranus) and zirconium. In addition, it was he who confirmed that all three metals were new, unique elements.

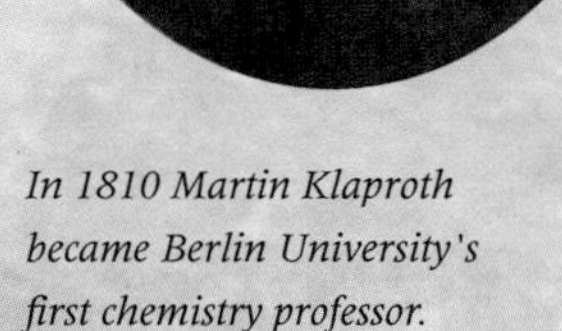

In 1810 Martin Klaproth became Berlin University's first chemistry professor.

39 A Standard Nomenclature

WITH THE SCIENCE OF CHEMISTRY STILL PERMEATED WITH THE CONFUSED LANGUAGE OF ALCHEMY, Antoine Lavoisier and his associates presented the first standard system of chemical nomenclature.

The language of chemistry is now familiar to most people, and many of its terms would be recognizable in everyday conversations—dioxide, carbonate, and sulfuric. When it was set out in the 1790s in the *Annales de Chimie*, a journal founded by Lavoisier, his wife Marie Anne, and various assistants, there was a logic attached that still persists today: a compound between a metal and a nonmetal has the suffix *-ide*, such as iron oxide. An acid was named after the non-oxygen portion, so oil of vitriol was renamed sulfur*ic* acid. However, acids of sulfur that contained less oxygen were qualified as sulfur*ous* acids. The compounds produced by acid had either *-ate* or *-ite* as the suffix. In other words nitric acid produced nitrates, while nitrous acid produced nitrites. As well as hydrogen and oxygen, Lavoisier promoted the word *gas* in place of "air." However, his name for nitrogen, *azote* (meaning "lifeless") did not catch on.

OFF WITH HIS HEAD

Antoine Lavoisier's contributions to science were recognized by the revolutionary regime that took over France in 1789, and even put to use. However, the scientist's great wealth was due to his position as tax collector for the hated king, and in 1794, Lavoisier's past caught up with him. He was guillotined at the age of 51.

Lavoisier's lab equipment is preserved at the Musée des Arts et Métiers in Paris.

40 Animal Electricity

AN UNINTENDED CONSEQUENCE OF A FROG DISSECTION SHOWED THAT ELECTRICITY WAS MORE THAN JUST SPARKS, but could be made to flow from one place to another. First thought to be confined to muscle, electric current was soon to be revealed as a more widespread phenomenon.

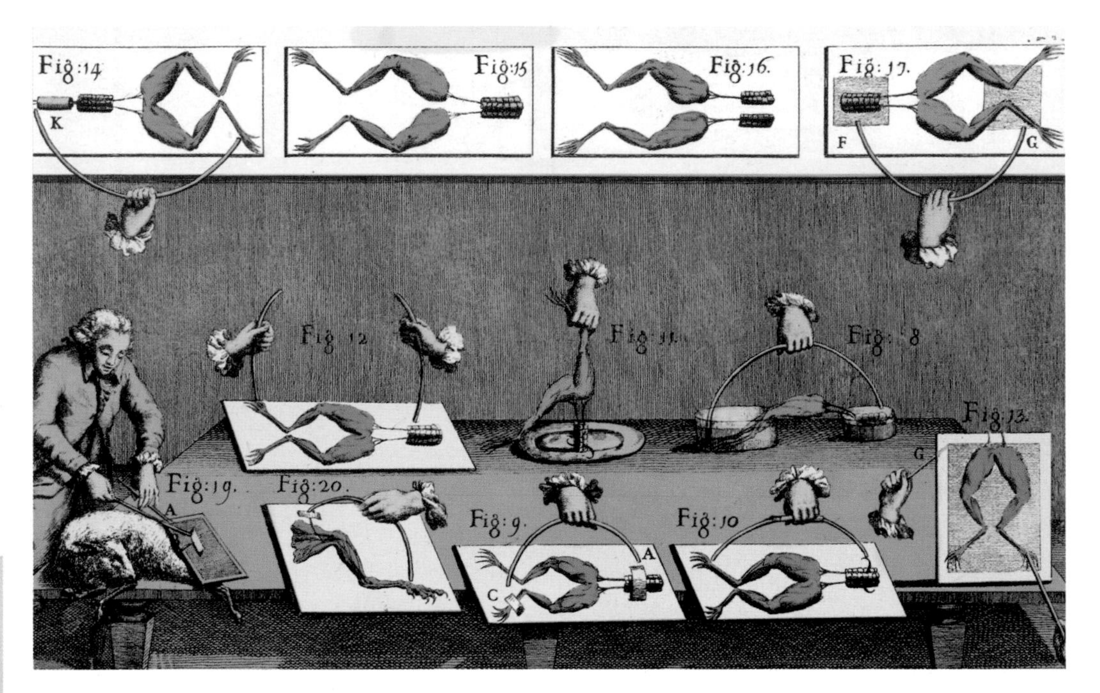

Galvani's drawings show the many ways he was able to manifest his "animal electricity."

GALVANIZING

Few people have words based on their names, but Luigi Galvani is one of them. The process of galvanization is named for him, which involves coating steel with a thin layer of zinc. If the zinc, and steel beneath, is scratched, an electrochemical reaction insures that the zinc fills the gap, not rust.

As is often the case in the furthering of scientific knowledge, the existence of electric currents was revealed purely by chance. In 1791, Italian anatomist Luigi Galvani was studying how nerves and muscles were connected. He hung up some disembodied frogs' legs on a wire fence to dry out. The fence was made of iron, while the hook was copper. The frogs' legs then twitched. (He later told a taller story about how he saw sparks flying when he and his assistant touched the nerves with a metal scalpel.)

Completed circuit

Galvani investigated further to find that he could repeat the phenomenon by using a curved wire—again made from copper and iron—to connect the exposed nerve to the end of the leg. Unknown to Galvani, he had formed a primitive circuit that allowed his "animal electricity" to flow along the nerve and into the muscle, causing it to contract. Galvani reported his technique worked equally well on larger mammalian subjects and he even showed that the human body could be used as part of the circuit. Obviously, Galvani believed that he had found some curious animated quality of animal tissue. Other scientists later showed that animal tissue was not required. In fact it worked much more effectively without.

41 Chemical Electricity

ALESSANDRO VOLTA COPIED GALVANI'S EXPERIMENTS. He found that the twitching leg was just the visible result of a electric force that formed between the two metals used.

As his name suggests Alessandro Volta is the man behind the volt, the unit used to measure the force that pushes electrical currents down wires, through nerves, and across the sky as lightning. The Italian earned this accolade in 1800 by inventing the voltaic pile, the first electrical battery.

Volta realized that Galvani's animal electricity was the result of a chemical reaction between two metals that was somehow causing electricity to flow from one to the other. For this to work, the circuit needed a salty liquid (from the frogs' legs) that separated the metals. In his pile, Volta placed disks of wood pulp soaked in brine between silver coins and pieces of zinc cut to a similar size. On its own the pile did nothing. However when wires running from top and bottom touched they completed the circuit, allowing electricity to flow, causing sparks or charging pieces of gold foil making them repel each other. The controllable energy of the Voltaic pile and later designs of battery would soon revolutionize chemical analysis.

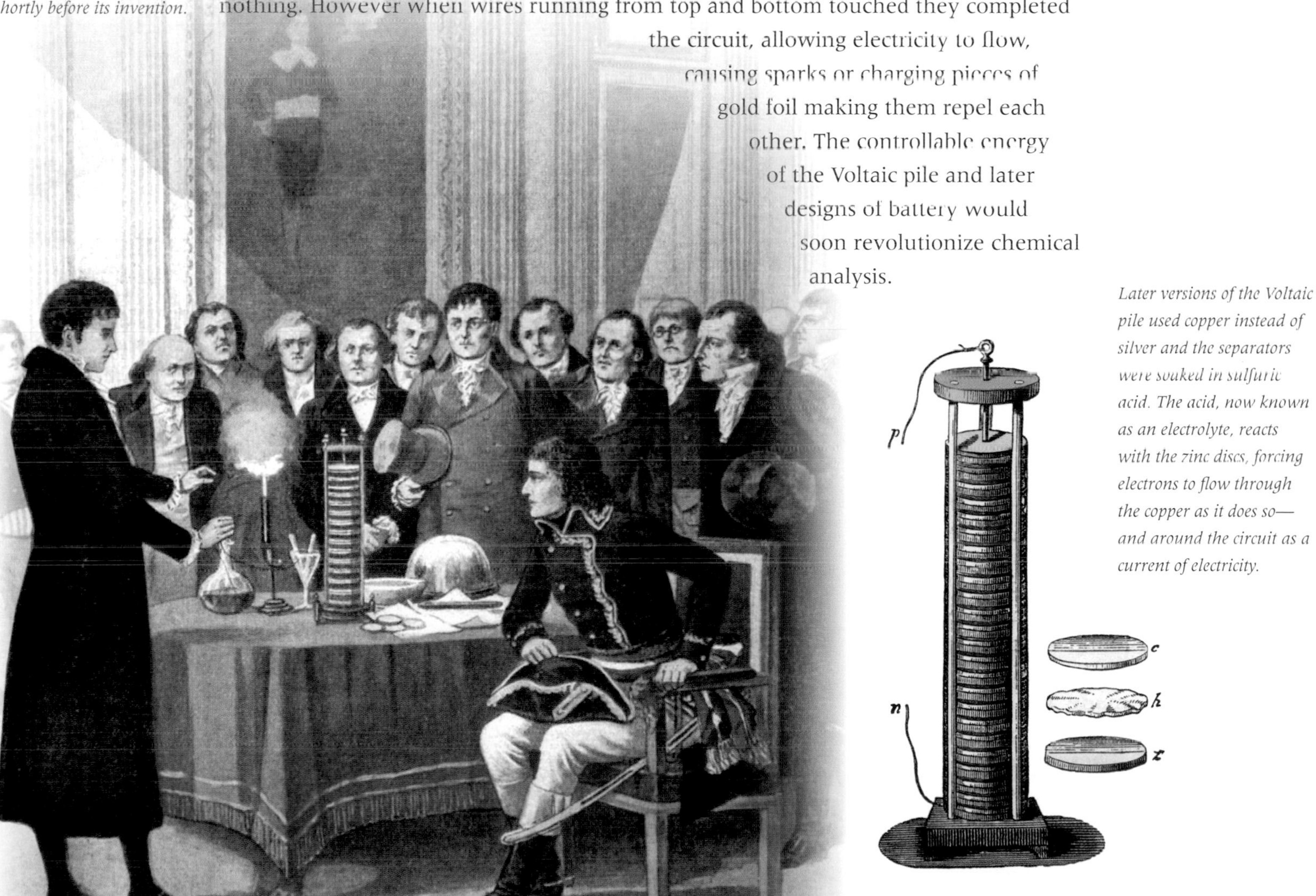

Alessandro Volta shows off his Voltaic pile to Napoleon who had invaded Italy shortly before its invention.

Later versions of the Voltaic pile used copper instead of silver and the separators were soaked in sulfuric acid. The acid, now known as an electrolyte, reacts with the zinc discs, forcing electrons to flow through the copper as it does so—and around the circuit as a current of electricity.

THE 19TH CENTURY: THE GREAT AGE OF SCIENCE (1800–1900)

42 Diffusing Gases

THE CONCEPT THAT AIR CONTAINED SEVERAL DISTINCT GASES HAD A FAR-REACHING EFFECT. The next big step was made by a meteorologist interested in how gases mixed and moved.

Englishman John Dalton's first interest was the weather. He recorded meteorological data for his whole adult life, and at the turn of the new century (while still in his 30s) Dalton applied this knowledge to the fundamental nature of gases. He revealed that even when mixed, individual gases diffused (spread out to fill a container) as separate entities. Here was the first evidence that the gases were made up of tiny, independent units that gave the gases their unique properties.

43 Atoms Reappear

JOHN DALTON'S WORK ON GASES LED HIM TO A STARTLING REALIZATION: THE GREEKS WERE RIGHT ALL ALONG. However, he was not referring to the theories of Aristotle but to the atoms of Democritus.

John Dalton prods a marsh bed to release bubbles of flammable marsh gas into jars. Marsh gas is better known today as methane.

John Dalton is credited as the man who reintroduced atoms to chemistry. In 1803, he proposed that gases were made up of minute and indestructible particles. He termed them atoms, just as they had been by Leucippus and his more famous student Democritus, when they first floated the idea 2,200 years before. To the Greeks, the atom was a philosophical construct, as much a product of thought as of nature.

Similarly, in 1738 Daniel Bernoulli was able to quantify the pressure exerted by a gas as a series of theoretical particles that were each exerting a tiny force pushing against the inside of the container. However, in Dalton's atomic theory the particles were thoroughly real objects, albeit too small to see directly. Despite their infinitesimal size, Dalton stated that it was atoms that gave gases their mass, and since a container of hydrogen weighed less than the same volume of oxygen, the atoms in each gas must be different.

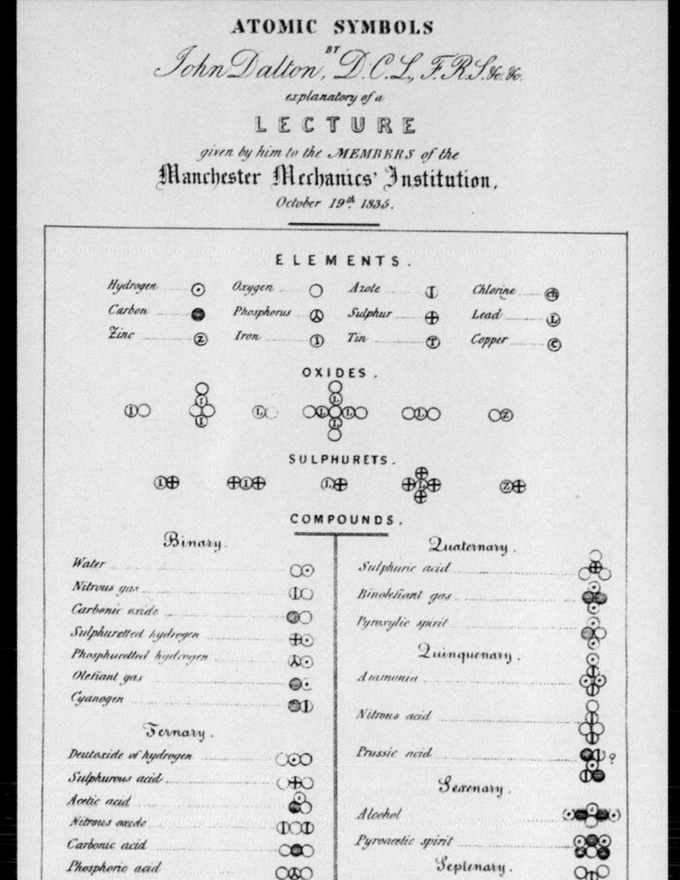

Dalton's table of elements drew on the past by using the alchemical device of assigning symbols to each element. However, it also reflected future developments by expressing simple compounds in terms of atoms bonded together in specific proportions and shapes.

Combinations and ratios

Dalton's experiments confirmed the intuition of others that elements combined in fixed ratios. He found that carbon and oxygen could combine in a ratio of 1:1 to make carbon monoxide, and also in the ratio 1:2 to form carbon dioxide (the "fixed air" of old). The ratios were always whole numbers, and the two elements did not combine in any other proportion. This discovery became the basis of the Law of Definite Proportions which was developed by Dalton and others of the time.

Dalton interpreted the ratios as atoms of different elements linking together. He imagined networks and clusters of atoms, which became known as *molecules* (a word coined a few years before). Dalton also used the ratios to calculate the atomic weights of different elements. He gave hydrogen a weight of 1 and then assigned values to the elements by comparing them with that gas. Dalton's reasoning was sound but his data was flawed, but nevertheless atomic weight (or atomic mass as it is now known) became the first parameter for organizing the known elements into a table.

44 Correct Proportions

John Dalton changed chemistry but it took the work of others to put it on the correct path. Crucially, Dalton had misjudged the nature of water, skewing the values for atomic weights.

Gay-Lussac studied the composition of atmosphere by balloon in 1804.

Joseph Louis Gay-Lussac is best remembered for a gas law named for him. Gay-Lussac's law states that the volume of a gas is proportional to its temperature—i.e. hot gas expands. In 1805 the Frenchman also showed that water was two parts hydrogen for every one of oxygen—H_2O not the 1:1 HO assumed by Dalton. Using Dalton's theory, an oxygen:hydrogen mass ratio of 16:2 gave oxygen an atomic mass of 8 (16/2). The actual figures achieved experimentally were slightly different but still wrong. Gay-Lussac showed that the atomic mass of oxygen should be 16 (as it stands today). With oxygen involved in so many compounds, correcting this mistake opened the door to charting the other elements.

45 Electrolysis

THE LINK BETWEEN ELECTRICITY AND CHEMICAL COMPOUNDS WAS MADE CLEAR IN THE EARLY 1800S. Electrolysis—splitting with electricity—became the latest tool in analyzing compounds.

DAVY LAMP

One of Humphry Davy's most enduring achievements was his safety lamp for miners. This device housed the naked flame inside a metal mesh which prevented any flames escaping from the lamp and igniting flammable gases contaminating the mine's air supply.

Humphry Davy was one of the first science celebrities and not without good reason. He began his scientific career early as assistant to Thomas Beddoes, who was studying the health effects of different gases in Bristol, England. While there, and against all advice, Davy inhaled nitrous oxide, which he found to be highly pleasurable—it is now known as laughing gas. Davy's gas caused a stir in the Bristol society of the day (the poet Samuel Coleridge became a fan) and soon became a popular lecture topic (given by Davy) at London's new Royal Institution, a public forum for the dissemination of knowledge.

Electrical experiments

As Davy was starting a career in London, news was spreading of Volta's electric pile. Davy and other members of the Royal Institution improved on its design and by 1807 they had built a large "battery" of silver-zinc cells. Davy directed its current at caustic potash and soda, two "earths" that had been included on Dalton's list of elements. Davy knew that electricity could split compounds and to his great joy, the melted potash (potassium hydroxide) broke down into a metal that immediately burst into flames. This was potassium, the first element Davy discovered. Caustic soda (sodium hydroxide) rendered sodium, and later electrolytic investigations added magnesium, calcium, boron, and barium to Davy's personal haul of elements.

The battery built by William Wollaston at the Royal Institution in 1807, the most powerful electrical source of its day, was used by Humphry Davy for electrolysis.

46 Halogens: The Salt-Formers

HUMPHRY DAVY'S EXPERIMENTS YIELDED MORE THAN JUST NEW METALS. He also showed that a pale green gas discovered by Carl Scheele 35 years previously was in fact a new element, the first of a potent family of substances called the halogens.

The first record of chlorine can be traced back to the 17th century in the work of Belgian alchemist Johannes van Helmont. However, the discovery of this acrid green gas is credited to Swede Carl Scheele 150 years later. Scheele produced it by chemically displacing it from hydrochloric acid—known to him as muriatic acid. Scheele believed that the gas contained oxygen, or fire-air as he knew it, and if that were removed a new element called *muriaticum* would be revealed. He failed to do so.

In 1810, Humphry Davy tried again, and declared that the gas was a new element, named chlorine for its greenish hue. Davy's conclusion also ended the notion that acids had to contain oxygen.

Sodium burns in the green fumes of chlorine gas to produce common salt, or sodium chloride to give it its chemical name, the very substance used for centuries to season food.

Salt-formers

Chlorine was a powerful substance but not an acidic one. Instead of turning litmus red (as an acid would) it bleached it white, a property of chlorine that has found many uses since. Chlorine also reacted readily with metals to form salty substances—chlorine and sodium formed common salt itself. The association with salts led some to suggest chlorine be called *halogen*, the "salt-former," in accordance with the naming of oxygen, nitrogen, and hydrogen. However, chlorine stuck.

In 1811, another of these salt-forming elements was isolated from the ash of burned seaweed. This metallic-gray solid sublimated into purple vapors, earning it the name iodine, from the Greek for "violet." Another was known to exist in fluoric acid, an extremely corrosive substance derived from fluorite. Fluorite had long been used as a flux in smelters and was named for the way it helped ores "flow" or melt. Humphry Davy suggested the unseen element be named fluorine (although it would take another 75 years to figure out how to isolate it). When liquid bromine was discovered in 1825, a family of elements was becoming clear, and halogen was adopted as their collective name. In the halogens, chemists were given a clear signal that elements could differ wildly in their physical properties but still share the same chemistry.

BROMINE

As one of just two liquid elements, and the only nonmetal one (mercury is the other), bromine holds a special status. The third halogen is named for the Greek word for "stench" because of the acid nature of its dark brown fumes. Dissolved bromides give the Dead Sea's water its high density (so swimmers float) while bromides also gave the rich red coloring to the confusingly named royal purple dyes that were worn by Roman nobility.

47 Avogadro's Law

LARGELY IGNORED IN HIS DAY, THE WORK OF AMEDEO AVOGADRO LIES AT THE HEART OF THE MODERN PERIODIC TABLE. The law attributed to the Italian states that a volume of gas, irrespective of its constituents, contains the same number of particles as another gas sample of the same volume.

With his meteorologist's hat still on, John Dalton showed that the total pressure exerted by a mixture of gases was the sum of the *partial pressures* of its constituents. Assuming temperature and volume are kept constant, the total pressure of the gas is proportional to the number of particles in the gas. Raising the partial pressure of one constituent gas while lowering that of another did not change the overall pressure. In 1811, Avogadro took this a little further and deduced that while the pressure of a gas was dependent on the number of particles, it was independent of the type of particle—and so all gases filling a fixed volume must contain the same number of particles, irrespective of their size and mass. Once fully recognized years later, this law became the key to unlocking chemical formulae and atomic weights.

AVOGADRO'S CONSTANT

Eggs come in dozens, but chemicals are counted in moles. Like a dozen, a mole is a set number, used to count very small things such as atoms and molecules. One mole contains about 602,214,129,270,000,000,000,000 particles. In terms of gas (any gas), one mole fills 22 liters (nearly 6 gallons). This number, generally shortened to 6.022×10^{23}, is called Avogadro's constant.

48 Introducing Symbols and Formulae

ONE OF THE LEADING ADVOCATES OF THE NEW ATOMISM WAS JÖNS JACOB BERZELIUS. The influence of this Swedish chemist can be seen running through modern chemistry, not least because of the way he chose to represent the simple compounds he was investigating.

By the 1820s, Jöns Jacob Berzelius had become one of the first chemists to employ Avogadro's law in their work. Like his predecessor, he realized that two gases reacting did not automatically result in fewer, albeit more complex molecules, such as single atoms combining into pairs, for example. Such a scenario would result in a drop in gas pressure. However, it was just as likely that the pressure of the gas was the same after

Berzelius is credited with the discovery of silicon, selenium, thorium, and cerium, while lithium was also first identified by an assistant working in his laboratory at Stockholm's Karolinska Institute.

the reaction as it was before. This showed that the number of particles was remaining constant. The simplest explanation for this was that a pair of atoms of one element (oxygen for example) was reacting with a pair of another (nitrogen) to form two pairs with atoms of each elements (nitric oxide).

Simple notation

Berzelius was among the first to see that reactions involved molecules of two or more atoms breaking apart and reforming with different constituents in different arrangements. He developed a simple notation system to express these complex changes. First he gave every element a symbol, generally the first letter of its name. So hydrogen became H, oxygen O, and nitrogen N. However, the names of many elements were not consistent in all languages, so Berzelius reverted to Latin for their symbols: iron took Fe from *ferrum*, sodium became Na from the mineral natron, while lead's Pb stems from the Latin word *plumbum*.

Compounds were represented as a formula, or combinations of these symbols. For example, nitric oxide was written as NO. A superscript number showed proportions, so water was written as H^2O—or it would have been if Berzelius had believed that was correct. He still thought water's formula was HO. The superscript was later changed to the subscript used today (H_2O), but Berzelius's basic shorthand is still in use. Today, it is extended to equations that show the molecules before and after a reaction.

ALCHEMICAL SYMBOLS

Berzelius's notation was not wholly innovative. Dalton and several of his predecessors had given symbols to the elements, as did the alchemists of old. These symbols were partly for convenience, partly a code, and partly magical figures that implied the paranormal character of each substance. There was no agreed system but alchemists often took their inspiration from astrology, with gold being symbolized as the Sun and silver as the Moon. Iron was linked with Mars, and that symbol is used today for maleness. Copper was controlled by Venus, and its sign is now used to signify female.

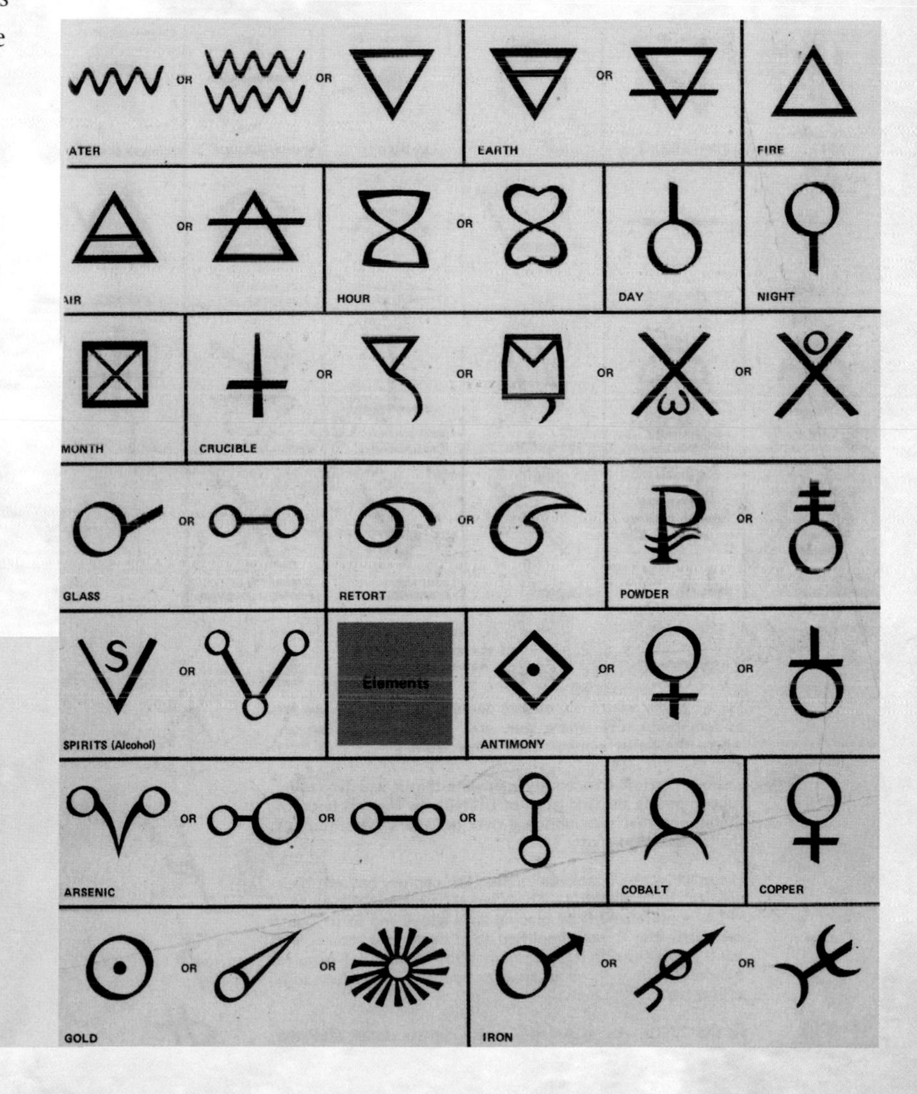

49 Electromagnetism

IN 1820 HANS CHRISTIAN ØRSTED DISCOVERED THAT ELECTRICITY AND MAGNETISM WERE CLOSELY CONNECTED. This discovery ultimately led to the technology that harnessed electrical power, but it also threw light on the force—later named electromagnetism—at work during chemical reactions.

PURIFYING ALUMINUM

Although Humphry Davy and others had made alloys of aluminum mixed with other metals, such as iron, Hans Christian Ørsted was the first person to make a sample of pure aluminum in 1825. He reduced it from aluminum chloride using pure potassium. His method, later improved by Friedrich Wöhler, was the main way of refining the metal until industrial electrolysis was introduced in the 1880s.

In some ways it was no accident that the Danish Ørsted made his discovery. He had a strong belief, in accordance with the philosophy of Immanuel Kant (recently deceased across the Baltic Sea), that scientific phenomena were all related as different facets of the same fundamental natural order. Nevertheless, the actual discovery for which he is remembered does appear to be have been by chance.

Ørsted was a lecturer at the University of Copenhagen, where he researched electricity and acoustics. During a lecture he noticed that when he ran a current from a voltaic pile through a wire it caused a compass to swing away from north and point torward the wire. When he switched the current off, the compass needle returned to the north. Although few of his audience took much interest, Ørsted immediately saw what it meant: the electrical current was making the wire— in all probability made of copper—into a temporary magnet. (Electromagnets work in precisely this way and are used as magnets that can be turned on and off.)

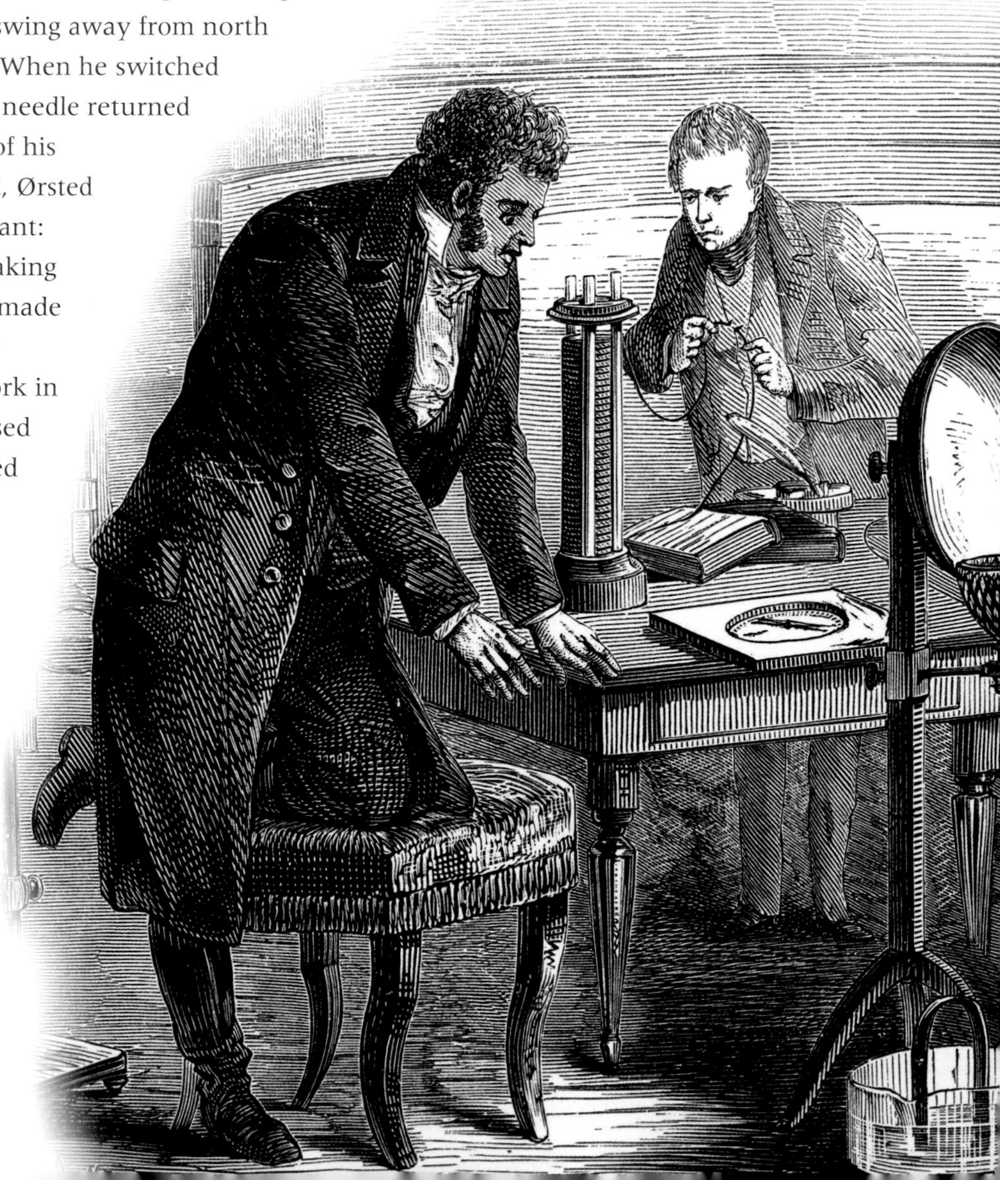

Hans Christian Ørsted checks his discovery as an assistant connects wires to an electric pile in the lecture theater in Copenhagen.

Radiating force

Ørsted thought that the magnetism was being given out from the wire, in the same way as heat or light is emitted. (It was a common conception of the time that electricity was some kind of liquid light, another intuition that the two phenomena were somehow linked.) Within a few months André-Marie Ampère had shown that currents running in opposite directions caused a repulsive magnetic force, while currents running in the same direction attracted each other.

Although he was unable to explain what linked the electricity with magnetism, Ørsted showed that within materials there was a two-way force at work that was transferring energy from one mass to another. In the following decades this pushing and pulling electromagnetism would be implicated as the force that drives chemistry.

50 Disproving the Vital Force

ONE OF THE LAST ANCIENT GREEK THEORIES TO COME TUMBLING DOWN WAS THAT OF THE VITAL FORCE. Once again, it took a bit of a luck.

ROMAN DOCTOR

Vitalism, the concept of a vital force, can be laid at the door of the Roman physician Galen. He was the surgeon to the gladiators at Pergamon (now in Turkey) and had plenty of opportunities to study living and dead tissue. He espoused that the body required more than a set of anatomical structures to be alive.

Galen lectures on anatomy in Rome.

Despite the many leaps made by chemistry during the century before, the scientists of the early 19th century still harbored a belief, a superstition that stretched back to ancient times, that the chemistry of life was not governed by the same rules as *inorganic* substances. A living body was run through with a vital force that animated otherwise lifeless material. Consequently, it was believed that the *organic* compounds that made up cells and tissue, could be analyzed (as many had by this stage) but could not be synthesized from inorganic materials. By way of proof, chemists showed how organic compounds "cooked" or denatured when heated, transforming in an irreversible process into inanimate, inorganic forms.

Friedrich Wöhler had been Jöns Jakob Berzelius's assistant and was a friend of Hans Christian Ørsted.

Making urea

In 1828, Friedrich Wöhler became the co-discoverer of beryllium, but he is better remembered for his attempts in the same year to make ammonium cyanate, a theoretical compound of nitrogen, carbon, and oxygen. He ended up with urea instead. Urea had been discovered 100 years before as the main constituent of mammal urine. Its molecule has the same constituent atoms as ammonium cyanate, and the latter appeared to spontaneously reform into the former. Wöhler had stumbled upon a way of making an organic compound from inorganic materials. Here was the first hint that the processes driving life were within the scope of chemistry.

51 Electrical Forces at Work

JÖNS JACOB BERZELIUS FIRST MADE THE LINK BETWEEN ELECTRICAL FORCES and the bonds between atoms. However, his initial interests were elsewhere.

Berzelius's laboratory is preserved at a museum devoted to the scientist in Stockholm, Sweden.

Berzelius's medical training coincided with the invention of the voltaic pile, and the young doctor soon began electrifying patients for therapeutic gain. This was not a success, but it led to another of the Swede's ground-breaking contributions to chemistry. Throughout his career, Berzelius promoted "a dualistic theory of chemical affinity," which stated that the reason certain atoms bonded (and not others) was because they were pulled together by opposite electrical charges. His idea fit well with evidence when it came to inorganic compounds but less so with organic substances, leading some to suggest that these bonded in different ways.

52 Ions or Radicals

ONE OF SCIENCE'S GREATEST FIGURES, MICHAEL FARADAY, NOW ENTERED THE DEBATE ABOUT THE ROLE OF ELECTROMAGNETISM IN CHEMISTRY. He suggested small but significant changes to Berzelius's theory.

Michael Faraday spent his early career in the shadow of another great scientist. He was Humphry Davy's assistant for more than a decade. In his own right, Faraday discovered benzene, later to be revealed as a fundamental organic compound. He also built on the work of Davy and William Hyde Wollaston to develop an embryonic electric motor in 1821—incurring the professional wrath of these two mentors in the process.

Faraday took a closer look at electrolysis, the analytical technique so championed by his boss. He revealed that the quantity of material decomposed by the electricity was proportional to the magnitude of the current. Although these concepts were yet to be

ELECTROMAGNETIC INDUCTION

Faraday is most famous for his 1831 discovery of induction, in which an electric current is induced in a wire or other conductor when it moves through a magnetic field. (Joseph Henry is credited with an independent discovery around the same time.) This phenomenon is the one harnessed by an electrical generator in a power plant, which converts the energy in motion into electrical currents.

Faraday presents his discovery of induction to the Royal Institution.

crystalized, this showed that energy carried by the electricity was being transferred to the compounds, causing them to split into simpler constituents.

Faraday (right) shows off some laboratory equipment to John Frederic Daniell, an English chemist who developed a powerful electrical cell which could produce more electricity then earlier piles and batteries.

Flow of energy

Faraday suggested that electricity flowed through a solution or molten liquid between the *electrodes* (a term coined by Faraday) by the movement of charged bodies he named *ions* (from the Greek for "wanderer"). He developed this idea further, putting forward the concept that molecules were held together by forces acting between ions of distinct and opposite charges. Berzelius disagreed that atoms were transforming into ions, preferring the idea that each atom had a positive and negative pole, which attracted (and repelled) the poles of others in the molecule.

Neologisms

Faraday needed a new language to describe his theory fully. He turned to his friend William Whewell for advice on other terminology. Whewell, who also introduced the term *scientist* as an alternative to "natural philosopher," suggested *anode* and *cathode* for the two electrodes, and *anion* and *cation* as their related ions—*an-* for positive charges, *cat-* for negative. These terms are still used today, which is a broad hint as to which theory proved to be the correct one!

53 Catalysts

SOME CHEMICAL INTERACTIONS RELY ON THE PRESENCE OF A THIRD SUBSTANCE, WITHOUT WHICH THE REACTION BARELY TAKES PLACE, IF AT ALL. Even today, this phenomenon is rather mysterious, and it was not even recognized as one until the 1830s, when the helpful substances were termed *catalysts*.

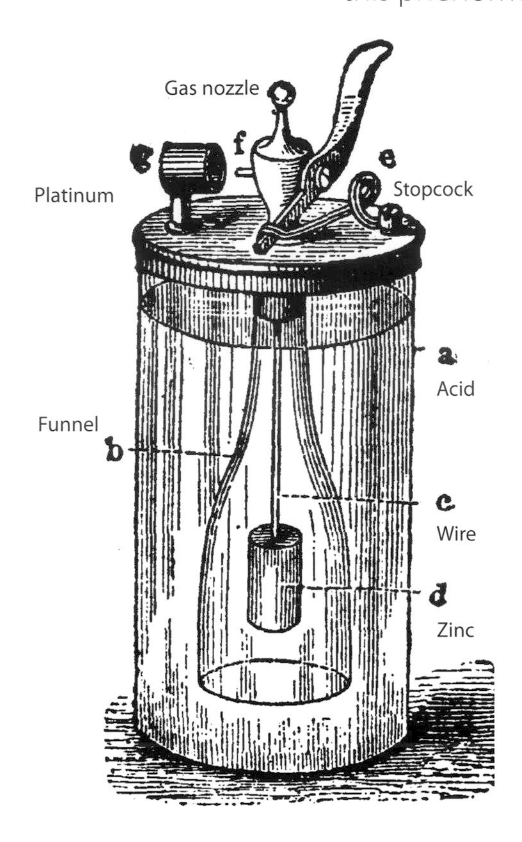

Döbereiner's lamp was a common lighter system for much of the 19th century, eventually superseded by the rise of hydrocarbon fuels such as paraffin, methane, and gasoline.

The concept of a catalyst was not new. In many ways yeast is the catalyst in the fermentation of wine and beer—at least until the rising alcohol levels kills it off. However, even by the 19th century the process of inorganic catalysis had yet to be tackled by chemistry, but that did not stop people making use of it.

In 1823 German chemist Johann Wolfgang Döbereiner designed a lighter system that produced a flame by burning hydrogen gas with the help of a catalyst. The gas was produced inside a glass bottle by reacting zinc with sulfuric acid. When it was released through a nozzle, the hydrogen passed through a mesh of platinum metal. Contact with the metal made the gas burn in the air—combining with oxygen to form water. Without the platinum, the hydrogen would simply escape, perhaps building up to a dangerous level that would cause an explosion. While the zinc and acid were depleted with use the lamp's platinum never ran out. Here was a true catalyst: without the platinum the reaction did not occur, but it was not used up in the reaction.

Breaking down barriers

As he had done so often in other areas of chemistry, the man who came up with the accepted name for this phenomenon was Jöns Jacob Berzelius. In 1836 he coined the word *catalysis* from the Greek word for "untie," alluding to the way the catalyst was able to break down barriers to the reaction in some way. A previous term used had been "contact action," which was also highly intuitive, since most catalysts are believed to work through a surface effect. The reactants are held temporarily on the surface of the catalyst. That may bring uncombined particles close enough to allow them to bond or it may stress a larger molecule in such a way that it splits up. The resulting substances then detach leaving the catalyst unchanged.

CATALYTIC CONVERTER

Platinum, along with rhodium, is the main catalyst in the catalytic converters fitted to cars. Its role is to convert corrosive and toxic gases into less harmful substances and reduce smog. So, as they pass through the cat, oxides of nitrogen are reduced to pure nitrogen while any carbon monoxide and unburned gasoline fumes are oxidized to form carbon dioxide and water. The cat requires a lead-free fuel, since the lead additives coat the catalytic mesh rendering it useless.

The valuable metal catalysts form a thin coating on a ceramic honeycomb core.

54 Chirality and Sidedness

A CENTURY BEFORE IT WAS POSSIBLE TO MAKE RUDIMENTARY IMAGES of even the largest molecules, Louis Pasteur found a way to literally throw light on the shapes of molecules.

Isomer is another word we can thank Swedish chemist Jöns Jacob Berzelius for. He coined it in 1830 in response to the work of his protégé Friedrich Wöhler who realized that cyanic acid and fulminic acid had the same elemental constituents but very different properties. The answer was that they were isomers: their identical ingredients were arranged into distinct structures.

Optical identification

In 1848, French chemist Louis Pasteur was studying tartaric acid, a naturally occurring crystalline substance that gives wine a sour taste. Sixteen years before, Jean Baptiste Biot had noticed that crystals of this acid derived from wine could rotate polarized light. (Polarized light is made up of waves that vibrate in the same plane—rotating the light changes the orientation of that plane.) Pasteur was interested in why tartaric acid synthesized in a laboratory did not do this despite being chemically identical to the natural acid. Studying the "synthetic" acid crystals under a microscope he found that there were two types, mirror images of each other. After sorting them out, he found that each group rotated polarized light in opposite directions. He had discovered chirality, in which isomers are mirror images of each other. Many natural substances, such as glucose, are chiral and yet are only produced in one form.

GERM THEORY
Louis Pasteur is best remembered for proving that disease and decay are caused by minute germs rather than being generated spontaneously by exposure to "bad" air. Pasteur took this theory and used it to develop the "pasteurization" technique, which sterilizes liquids such as wine, milk, and juices, using a intense but short blast of heat. This kills the germs but does not alter flavors too much.

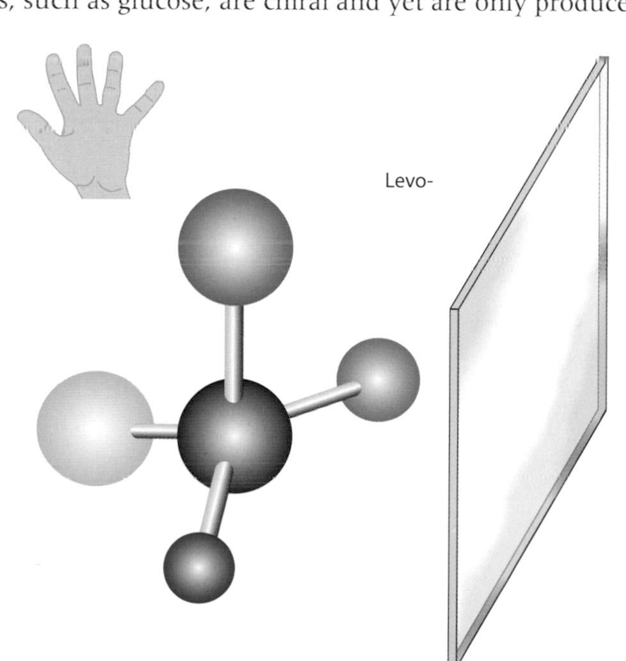

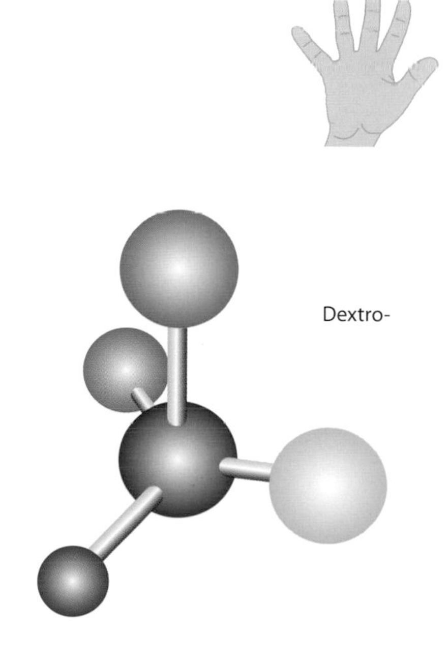

Chiral isomers, or enantiomers, are characterized as being dextro- or levo- which relates to their rotation of light to the right and left respectively. Like a pair of hands, the enantiomers of a substance cannot be superimposed on one another, no matter how they are rotated.

55 Valence and Molecules

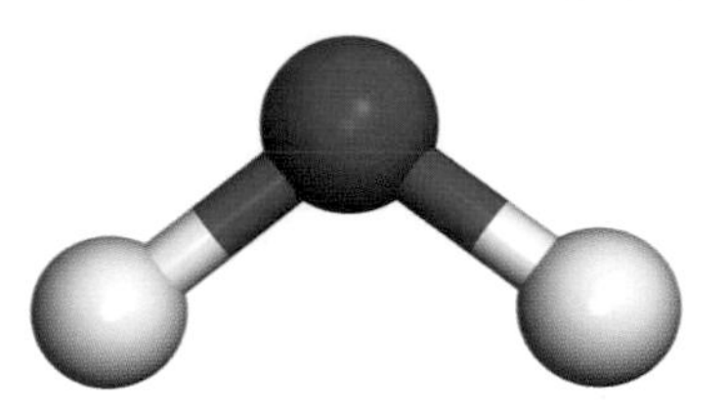

This molecule of water shows that oxygen (red) has a valence of two, while each hydrogen (white) has a valence of one.

INTEREST IN THE NEW FIELD OF ORGANIC CHEMISTRY RESULTED IN A CONCEPT that had a significant effect on our understanding of atoms and bonding.

Edward Frankland's career began as an apprentice druggist, preparing medicines for pharmacies in Lancaster in northern England. However, good fortune and better contacts saw him upgrade to an assistant in the Geological Society's lab in London. His employer introduced him to the German Robert Bunsen, not yet at the height of his fame but nevertheless influential. Frankland studied at Marburg under Bunsen for two years, which set him on his way to becoming one of the first chemistry professors at the University of Manchester in 1853.

In the 1850s, Frankland discovered that he could combine zinc and other metals with organic compounds, creating substances he dubbed organometallics. He found that a fixed amount of metal was used up in each synthesis, but this varied from metal to metal. This showed that the atoms of metallic elements did not all bond in the same way, but had a certain "combining value." This value was later named valence and indicated the maximum number of connections, or bonds, an atom could form within a molecule. Hydrogen and chlorine had a valence of one, while oxygen's was two. Most intriguing was carbon, present in all organic compounds, which had a valence of four.

VARIOUS VALENCES

The unusual atomic structure of the transition metals, the set of elements that fills the center of the periodic table, gives them a variable valence. For example manganese atoms can form up to seven bonds! All this variety gives the compounds of transition metals (below) a wide range of colors.

56 The Bunsen Burner

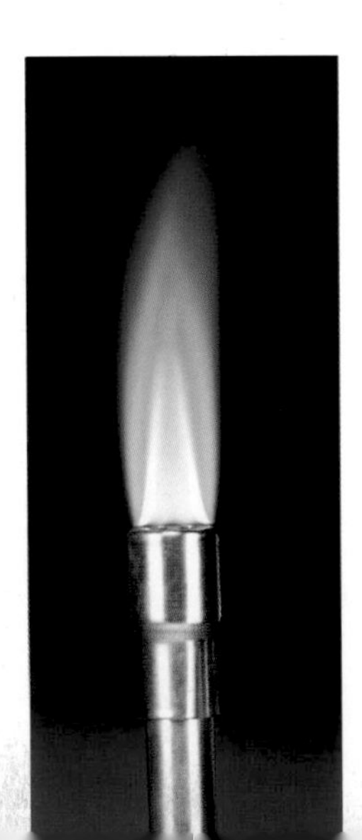

WHEN ROBERT BUNSEN TOOK UP A NEW POST IN HEIDELBERG HE WAS INSTALLED IN A HALF-BUILT LABORATORY supplied with coal gas, a flammable and toxic gas (due to the carbon monoxide) released by roasting coal.

Bunsen wanted the gas to be a source of light and heat, and he submitted a design for a burner that supplied both to the university's engineer Peter Desega. Air holes near the base ensured that gas and air were thoroughly mixed before igniting, producing a tall, bright yellow flame. Opening a collar at the base increased the air flow making a blue, hot flame (and minimum soot) ideal for heating glassware. Chemistry students began using Bunsen's burner in 1855 and have been doing so ever since.

The hottest flame of the Bunsen burner has a characteristic blue cone within. Modern burners are supplied with methane or propane, less toxic and hotter burning than the original coal gas.

The Geissler Tube

ALTHOUGH IT WAS LITTLE MORE THAN A ENTERTAINING CURIOSITY, THIS GLOWING DEVICE not only made a link between light, electricity, and atoms but was also the first step toward radio, television, and computer technology.

The Geissler (or Geißler) tube is named for its inventor Heinrich Geissler, an instrument-maker at the University of Bonn, who, in the late 1850s, was tasked by Julius Plücker to manufacture a glass tube with a near vacuum inside. Geissler was a skilled glassblower but his major contribution was the mercury-displacement pump that drew air out of the tube and blocked any replacement air from rushing back in.

Plücker removed as much gas as possible from the tube and passed through it the largest electric current that he could muster from the batteries of the day. The current made the gas glow faintly, a foretaste of the neon lights of the early 20th century. Plücker recorded that the luminous discharge was attracted to a magnet and that each gas glowed with a distinct color, two phenomena that would soon give scientists access to the inner workings of the atom.

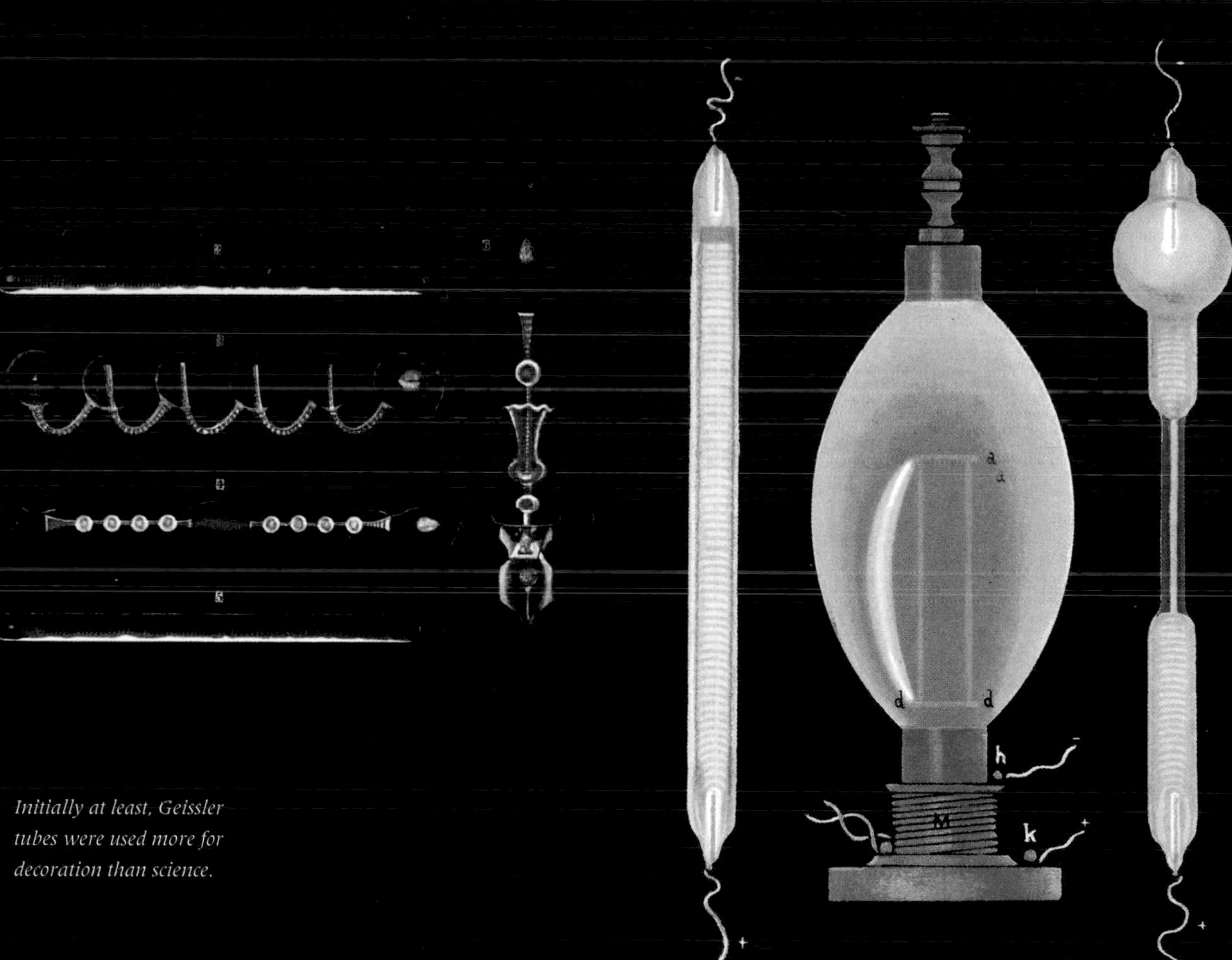

Initially at least, Geissler tubes were used more for decoration than science.

58 The First Plastic

TO TODAY'S EARS THE WORD PLASTIC IS FAR FROM EXOTIC. THE MATERIAL IS NOW SO UBIQUITOUS IT IS USED FOR JUST ABOUT EVERYTHING MEANT TO BE INEXPENSIVE OR DISPOSABLE. However, in the 19th century *plastic* referred to a remarkable property that allowed a substance to retain almost any form that it was shaped into.

Two barrettes made from Parkesine have survived from 1860. Pigments were mixed with the soft plastic before it was molded.

Early plastic materials were refined from natural products. Horn and whalebone were used for strong yet flexible items; shellac was a moldable resin secreted by an Indian plant-sucking bug; and natural rubber was made from milky plant secretions. Then in 1856 Englishman Alexander Parkes synthesized a substance that had all these properties. He called it Parkesine and made it by applying nitric acid to cellulose, the fibrous material in all plants, especially wood (and by extension paper). This had the effect of making the long molecules of cellulose join together with cross links to form a rigid solid. Parkesine became softer when heated and so could be molded into any shape. Parkesine was not a commercial success but had been refined into celluloid by 1870. This later material found fame as the long, flexible strips that put the "film" in photography and cinema.

59 Carbon Chemistry

THE DISCOVERY OF ATOMIC VALENCE, WITH CARBON HAVING A VALUE OF FOUR, WAS THE BREAKTHROUGH THAT UNLOCKED THE SECRETS OF ORGANIC CHEMISTRY through the work of Friedrich August Kekulé in 1858. However, this came against the backdrop of many theories developed by others over the previous decade.

Friedrich Kekulé reported that he dreamt of carbon chains while dozing on a London bus.

Simple combustion told chemists that organic compounds were hydrocarbons, compounds chiefly comprised of carbon and hydrogen that produced carbon dioxide and water when burned. However, it was unknown how these elements, and others like oxygen and nitrogen, associated with each other. Friedrich Wöhler had shown that organic compounds must share at least some of the properties of inorganic ones. He and his colleague Justus Leibig (the inventor of the Leibig condenser used in laboratory distillations) had also shown that the arrangement of atoms, the shape of the molecule they formed on the smallest scale, was central to the properties they exhibit on the large scale. Their French contemporary Jean-Baptiste Dumas proposed that carbon and hydrogen clusters, or radicals, formed the core of the compounds, held together by Berzelius's polar charges. Soon the ethylene (two carbons) and methyl (one carbon) radicals had been identified.

However, while Frankland's revelation of valence arrived through organometallics, Dumas's theory was put under strain by *organohalides*. By chance—and Faraday and others confirmed it—Dumas found that chlorine (and other halogens) could take the place of hydrogen in the organic compound. According to Berzelius's dualistic affinity theory, hydrogen was one of the positive elements, while chlorine was a negative one. How could a radical swap one for the other? Berzelius rather desperately suggested that the chlorine "copulated" with the radical, altering its shape and bonding to a different part. However, this was discounted since it equated to isomerism, and such a change of form would have resulted in a change of properties that was not apparent in the new organohalide.

The oil industry began in earnest at Titusville, Pennsylvania, in 1859, at first as a source of fuel and later as raw materials for the chemicals industry.

NAMING CONVENTIONS

Carbon chains are named according to the number of carbons in the longest section. Side branches, or alkyl groups, are also named according to the number of carbons: For example 2-methylbutane has a methyl sticking out of the second carbon of the butane chain.

Prefix	No of Carbons
Meth-	1
Eth-	2
Prop-	3
But-	4
Pent-	5
Hex-	6
Hept-	7
Oct-	8
Non-	9
Dec-	10

Tetravalent carbon

Then Friedrich Kekulé changed the game. Every carbon was bonded to a maximum of four other atoms. He realized that meant that carbon atoms could form into long, branching chains and rings. Such structures formed "a skeleton" around which other atoms, hydrogen chief among them, were arranged. Kekulé's illustrations of his theory have atoms connected by indistinct sausage-shaped zones—somewhat prescient of our current fuzzy view of atomic bonds.

Thanks to Kekulé, organic compounds follow a pattern: methane is a carbon and four hydrogens, while ethane is two carbons linked together and three hydrogens on each of them, followed by propane, butane, pentane... and a lot more besides.

60 Spectroscopy

It had frequently been noted that the flames of burning salts and other compounds had colors specific to the metal elements within. The study of these colors led to a whole new science.

The flame test was one of the first ports of call in a voyage of chemical discovery and an effective way of differentiating all too similar white salts. An orange flame indicated the presence of sodium, while lilac showed the salt contained potassium. The eerie glows from a Geissler tube added further weight to the idea that elements had some kind of color signatures, and it was to this cause that Robert Bunsen applied his new gas burner.

The Bunsen burner produced a clean and constant source of intense heat, with a faint blueish flame that did not interfere too much with the color of the burning sample. Nevertheless, Bunsen had trouble filtering out the true color of the sample. In stepped Gustav Kirchhoff, a fellow academic at Heidelberg. He suggested that the light be split into its constituent colors using a prism, as Isaac Newton had done 200 years before in his ground-breaking work on optics and the spectrum.

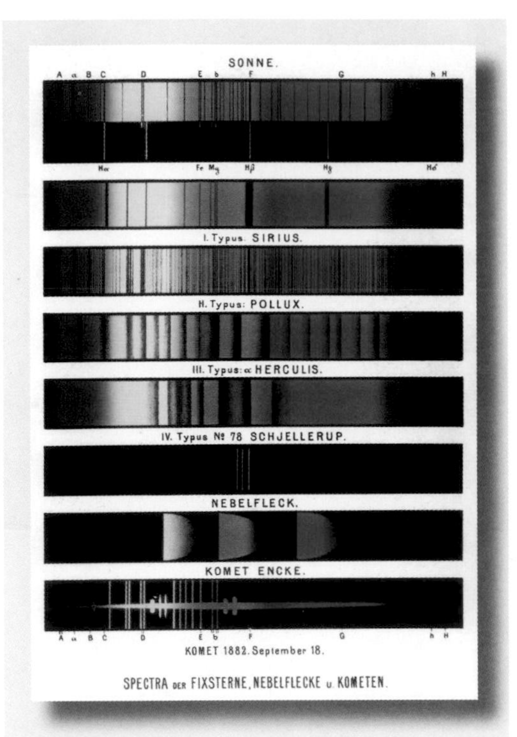

ABSORPTION AND EMISSION

Spectroscopy is underwritten by Kirschhoff's three laws: 1) Hot solids produce a full spectrum of colors (white light); 2 Hot gas (like a flame) glows with a specific set of colors (its emission spectrum); 3) Cold gas absorbs specific colors from white light leaving dark lines in the full spectrum (as seen above). These absorption spectra show us which elements form the gas and dust of space.

Spectrometer

Joseph von Fraunhofer had invented such a device for analyzing starlight in 1814. He had found that the full spectrum of colors was filled with dark lines—where certain colors were missing. Bunsen and Kirchhoff's spectrometer showed that the color of flames was not a full spectrum but comprised a handful of colors, which formed a series of faint lines.

Bunsen and Kirchhoff's spectrometer focused the flame's light onto a central prism, which directed light into a separate eyepiece.

The German scientists showed that each element emitted a specific family, or spectrum, of colors that could be used to identify them. In 1859 they used their spectrometer to identify the presence of two new metal elements, first cesium and them rubidium, both of which shared similarities with sodium and potassium. Spectroscopy showed that light was a feature common to all elements not an element in itself.

61 The Karlsruhe Congress

IN 1860, AUGUST KEKULÉ AND A FEW COLLEAGUES CALLED A CONFERENCE FOR THE WORLD'S CHEMISTS. The main topic of conversation at the meeting at Karlsruhe in southern Germany was how best to organize the known elements and to agree on a system of atomic masses.

For three days in September 1860, anyone who was anyone in the world of chemistry was at the Karlsruhe Congress, the first international chemistry conference. All the delegates (except for one Mexican) hailed from European universities, and included bigwigs such as Dumas and Bunsen. One lesser known attendee was Italian Stanislao Cannizzaro, who promoted the work of his late countryman Amedeo Avogadro as a method for calculating atomic masses. Cannizzaro was largely ignored until the last day, when his pamphlet reached the hands of a professor from Russia's St. Petersburg university, one Dmitri Mendeleev. Before the decade was out, Mendeleev would use the consistent atomic masses offered by Avogadro's law in order to build the periodic table itself.

62 Observing Helium

SUNLIGHT IS WHITE, A COMBINATION OF ALL THE COLORS OF THE RAINBOW. However, the light from the corona, the ring of gas that surrounds our nearest star, revealed more data when viewed through a spectroscope, including a new and, as it turned out, somewhat unusual element.

Non-flammable helium's low density makes it a safe alternative to hydrogen for use in modern airships.

Looking at the sun through a spectroscope (telescope and spectrometer combined) is only safe during an eclipse, when the blinding light is obscured leaving only the corona in view. In 1868, Pierre Janssen and Norman Lockyer both saw a distinct yellow line in coronal light and realized it was emitted by a new element, duly named helium for *helios*, "sun" in Greek.

63 The Periodic Table

DMITRI MENDELEEV IS HAILED AS THE FATHER OF THE PERIODIC TABLE, THE CENTRAL DOCUMENT OF CHEMISTRY. This was by no means the first attempt to classify the elements, and his work was naturally informed by the failings of others, but it is Mendeleev's table that has stood the test of time.

The most obvious way of organizing the elements is by atomic weight (now referred to as atomic mass). This is how John Dalton did it way back in 1803. However, by the 1860s there were more values for atomic weights than there were known elements—56 in 1863, rising to 64 listed on Mendeleev's first table in 1869. With a single method for calculating atomic weight yet to be agreed, cataloging chemists looked for other patterns within the chemical properties of elements. This was an uphill task since, as we now know, many elements remained undiscovered so obscuring the view.

Dmitri Mendeleev, complete with his trademark beard, at work in his St. Petersburg study in the late 1890s.

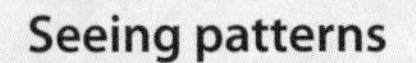

Seeing patterns

Johann Döbereiner (of catalytic lamp fame) had found five triads among the jumble. Lithium, sodium, and potassium was one, the three known halogens another. In 1865 English chemist John Newland noticed a "law of octaves" in which the chemical properties of elements appeared to fall into seven groups. This even tallied with atomic weights, which were firming up by now thanks to the work of Cannizzarro at Karlsruhe. Newland set out a table in which every eighth element was positioned below a predecessor that shared its properties. This arrangement worked well for Döbereiner's triads but the incomplete list of elements meant that Newland had to leave spaces elsewhere. He was rather inconsistent in his approach to this which left him open to ridicule from the chemistry community.

Typische Elemente							
			K = 39	Rb = 85	Cs = 133	—	—
			Ca = 40	Sr = 87	Ba = 137	—	—
			—	?Yt = 88?	?Di = 138?	Er = 178?	—
			Ti = 48?	Zr = 90	Ce = 140?	?La = 180?	Th = 231
			V = 51	Nb = 94	—	Ta = 182	—
			Cr = 52	Mo = 96	—	W = 184	U = 240
			Mn = 55	—	—	—	—
			Fe = 56	Ru = 104	—	Os = 195?	—
			Co = 59	Rh = 104	—	Ir = 197	—
			Ni = 59	Pd = 106	—	Pt = 198?	—
H = 1	Li = 7	Na = 23	Cu = 63	Ag = 108	—	Au = 199?	—
	Be = 9,4	Mg = 24	Zn = 65	Cd = 112	—	Hg = 200	—
	B = 11	Al = 27,3	—	In = 113	—	Tl = 204	—
	C = 12	Si = 28	—	Sn = 118	—	Pb = 207	—
	N = 14	P = 31	As = 75	Sb = 122	—	Bi = 208	—
	O = 16	S = 32	Se = 78	Te = 125?	—	—	—
	F = 19	Cl = 35,5	Br = 80	J = 127	—	—	—

Mendeleev's original periodic table from 1869. In this early version the periods were arranged in columns. A revised form in 1871 reoriented the table so periods ran across and elements with similar properties were grouped into columns.

Applying valence

Mendeleev added valence to Newland's scheme, which gave him another parameter to position the known elements and leave suitable blanks. This was one significant difference between the two approaches. Unlike Newland, Mendeleev argued firmly (and with great success) that the blanks in his table represented undiscovered elements.

Reportedly inspired by the game Patience (or Solitaire), Mendeleev made cards for each element so he could arrange them in different ways. His final version had a series of "periods," so named because they exhibited the same repeating rhythm of properties that Newland had noticed. The first period contained just hydrogen because Mendeleev did not accept the existence of helium (the next heaviest element) until 1902. Therefore, the next element available was lithium. Like hydrogen this has a valence of one and so it became the first member of the next period. Beryllium, boron, and carbon followed, each with a higher weight and valence than their predecessor. The next three elements, nitrogen to fluorine, had decreasing valences, but also nonmetallic properties very different to the metals at the start of the period. The next element was sodium. Metallic and with a valence of 1, it began the next period. Mendeleev's table worked so well because, without knowing how, he was reflecting the fundamental structure of atoms which would only be revealed 40 years later.

PREDICTING PROPERTIES

In a bold move, Mendeleev used his table to predict the properties of as yet undiscovered elements. Two spaces fell between tin and arsenic. The first Mendeleev called eka-aluminum, after the metal that was located in the period above, and predicted its valence and approximate density and melting point. He did the same with its neighbor, which he called eka-silicon. By 1885 the Russian had been proved right; eka-aluminum was renamed gallium, while eka-silicon is now germanium.

64 Cathode Rays

TECHNOLOGY UPGRADES TO GEISSLER TUBES RESULTED IN RAYS THAT DID not merely glow but could cast a shadow. The mysterious rays shared features with electricity, magnets, and light. But which were they?

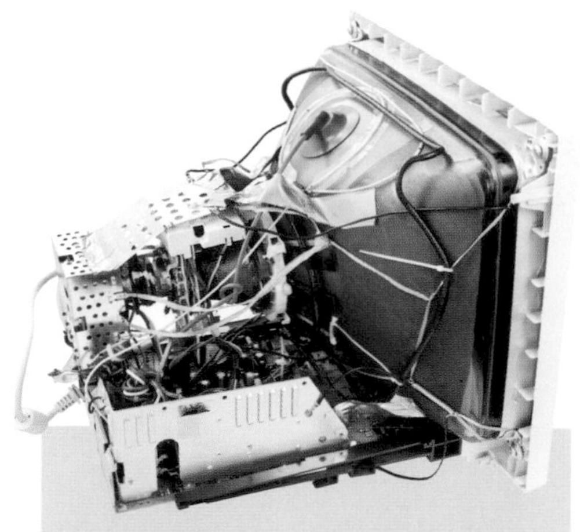

The cathode-ray tube in an old-fashioned TV works in the same way as a Crookes tube.

In the early 1870s, English physicist William Crookes developed a better, more powerful design of gas-discharge tube, akin to the Geissler tubes invented in the 1850s. Crookes invented a new vacuum pump that reduced the amount of gas in the tube to a level 10,000 times less than that of a Geissler tube. In addition, the voltages that Crookes was able to apply to the gas far outstripped the power of Geissler's device.

A dark space

The current ran though the rarefied gas between two electrodes, the anode and cathode. However, Crookes found that the combination of less gas and more electricity did not just result in a brighter glow. The area close to the cathode remained dark, with the glow increasing as it neared the anode. By then it had formed a beam that flew past the anode, forming an eerie fluorescence at the end of the tube.

Common sense dictated that the beam of rays was emitted from the cathode rather than the anode, so the mysterious light, best seen in a darkened room, was named cathode rays. Cathode rays were found to have a definite direction and were not radiating out in all directions from the cathode, like heat or light would. The rays were also bent by magnetic fields. They even appeared to spin paddle wheels fitted inside the tube, adding to a dazzling array of properties that would soon reveal the rays as coming from the insides of atoms themselves.

WATCHING THE TUBE

It is often said by the elders among us that the younger generation watch too much of the "tube." Of course they are referring to the TV, which until recently, before the advent of "flatscreens," formed images with cathode rays. The cathode-ray tube was a little more advanced than a Crookes tube. A flickering beam sweeps across the back of the screen under the direction of an electromagnet. The inside of the screen is coated with phosphors, materials that convert the largely invisible cathode rays into dots of light (and color) that make up the moving image.

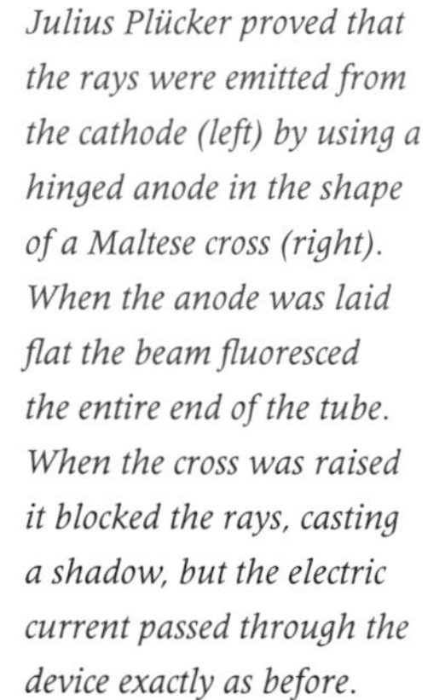

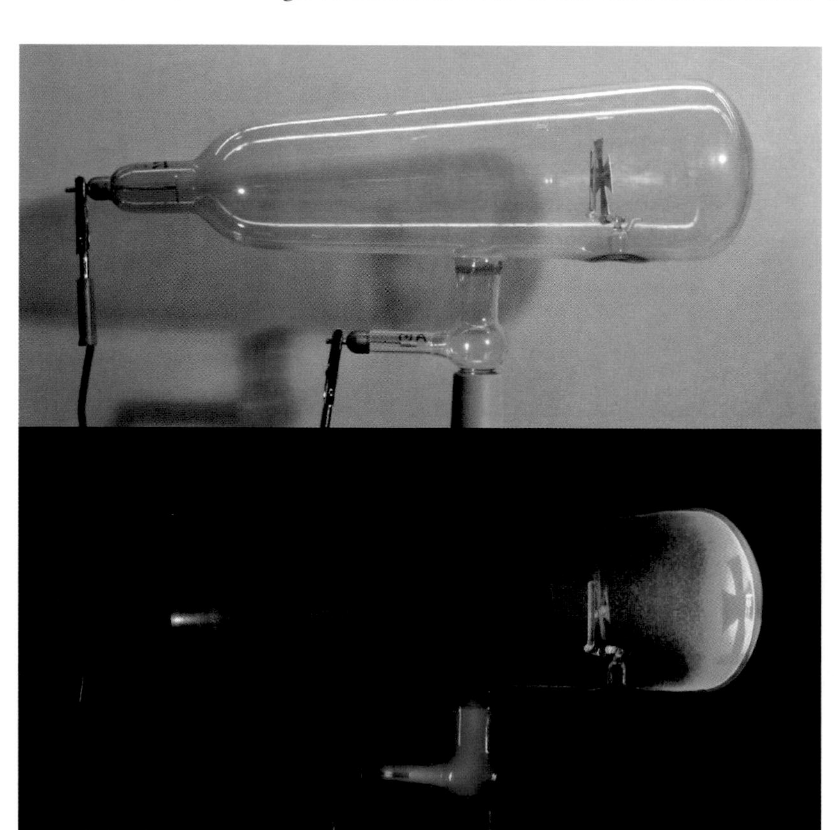

Julius Plücker proved that the rays were emitted from the cathode (left) by using a hinged anode in the shape of a Maltese cross (right). When the anode was laid flat the beam fluoresced the entire end of the tube. When the cross was raised it blocked the rays, casting a shadow, but the electric current passed through the device exactly as before.

65 Semiconductors

MATERIALS THAT ACT AS BOTH CONDUCTORS AND INSULATORS FORM THE BASIS OF OUR MODERN, DIGITIZED CIVILIZATION. The moment when the phenomenon was discovered remains largely forgotten.

Ferdinand Braun made his discovery of semiconduction while working as a high-school teacher.

In the 1820s, German physicist Georg Ohm set out his eponymous law which stated that the current through a conductor was proportional to the voltage pushing it along. The coefficient linking the two values was the resistance. Until the 1870s, it was thought that resistance was a largely constant value. When it was so high that no current ran at all, that material was termed an insulator and not of concern to Ohm's law.

Crystal conductors

In 1876, a young graduate Ferdinand Braun presented his discovery of "deviations from Ohm's law" to the Natural Society of Leipzig. Braun had found unusual properties in crystals of galena, a naturally occurring form of lead sulfide, and in other minerals. He was not the first person to test their conductivity. However, he did so using very fine, needle-shaped electrodes that made a strong connection with the crystals. Braun noted two phenomena. Firstly some crystals acted as conductors when the electrodes were applied, but then became insulators when the direction of the current was reversed. Secondly, crystals switched from being insulators to conductors as the voltage was raised, and the boost in current observed was not in line with the predictions of Ohm's law.

Braun went on to an illustrious career, winning the Nobel Prize alongside Guglielmo Marconi in 1909, for his contributions to radio technology. He is however best remembered for inventing the cathode-ray oscilloscope, a forerunner of the first television sets.

FROM ELECTRICS TO ELECTRONICS

It took another 60 years to figure out what was happening in the atoms of silicon and other semiconductors. When that challenge was overcome there was a sudden boom in the electronics industry. However, Braun's first discovery had already been put to use in a rectifier, a device that only allowed current to move in one direction, blocking flow in the other. Rectifiers are used to transform alternating currents into direct ones, and were an essential component in the first radio sets. The second phenomenon revealed by Braun was that the electrical properties of semiconductors could be switched on and off, which is exactly how they are used in electronic circuitry such as the microprocessors in computers.

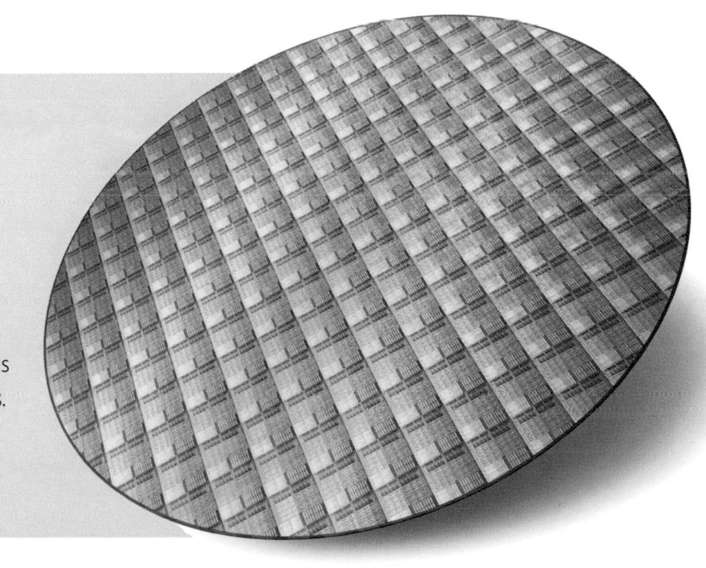

Dozens of integrated circuits, or microchips, are etched on to the surface of a silicon wafer, a single, flawless crystal of silicon.

66 Activation Energy

THE MAJORITY OF CHEMICAL REACTIONS REQUIRE AT LEAST A SPARK TO SET THEM OFF. While a few reactions, such as sodium in water or magnesium in strong acid, seemed to occur spontaneously, most needed a helping hand to get them going. In 1889, a Swedish chemist came up with a way of explaining these "energy barriers."

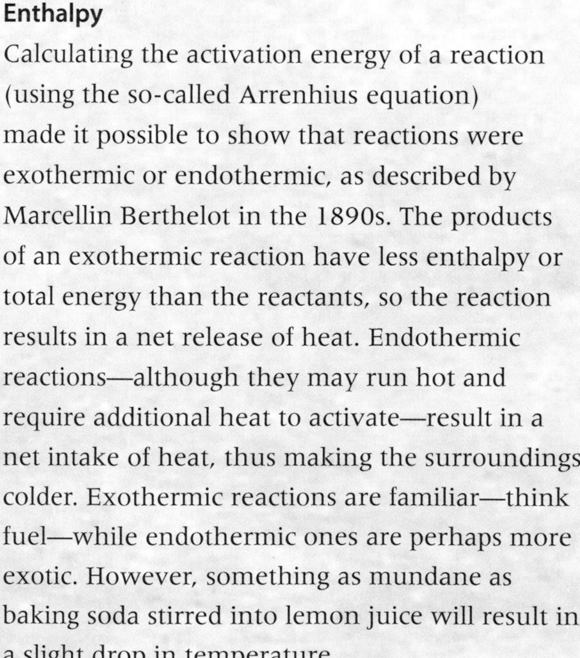

Svante Arrenhius at work in his Stockholm laboratory. The Swede is remembered for correctly theorizing that acidity was caused by chemicals dissociating into hyper-reactive hydrogen ions. This was later codified by others into pH, a measure of the "potential hydrogen."

Svante Arrenhius is another towering figure in chemistry. He was one of the first scientists to quantify the warming effect of carbon dioxide in the atmosphere, later to be termed the greenhouse effect, now thought to be a major driver of climate change.

Energy barrier

In 1889, Arrenhius used the term "activation energy" to describe the energy barrier that two reactants had to overcome in order to form products. The higher this activation energy the less likely the reaction would take place. Temperature is a measure of the energy in a substance that gives an average for all the particles in it. Even in cold samples a very few molecules might have the requisite energy to react so the process could run, but only slowly. Heating the sample adds energy to it and increases the number of molecules with enough energy to get over the barrier and form products.

Enthalpy

Calculating the activation energy of a reaction (using the so-called Arrenhius equation) made it possible to show that reactions were exothermic or endothermic, as described by Marcellin Berthelot in the 1890s. The products of an exothermic reaction have less enthalpy or total energy than the reactants, so the reaction results in a net release of heat. Endothermic reactions—although they may run hot and require additional heat to activate—result in a net intake of heat, thus making the surroundings colder. Exothermic reactions are familiar—think fuel—while endothermic ones are perhaps more exotic. However, something as mundane as baking soda stirred into lemon juice will result in a slight drop in temperature.

The thermite welding process displaces the iron in ore with aluminum to produce pure iron and a great deal of heat. The activation energy of this reaction is high—almost 2,000°C—but it gives out a great deal more energy than is taken in.

67 X Rays

IT HAD LONG BEEN SUSPECTED THAT THERE WAS MORE TO CATHODE RAYS than the visible glow. Several investigators had found that an invisible beam was clouding photographic paper. One of them noted its presence with a quizzical "X" and the name stuck.

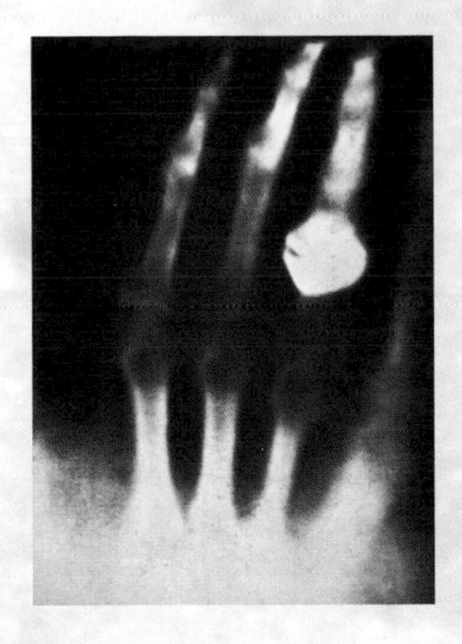

The ring on Anna Röntgen's finger stands out with her bones in this first X-ray photograph from 1895. On seeing the picture, the subject is reported to have said, "I have seen my own death."

The researcher responsible was the German Wilhelm Röntgen, who recorded the rays in 1895. The precise discovery is shrouded in uncertainty but it is thought that Röntgen had wrapped up a Crookes tube (perhaps modified with an aluminum "Lenard" window which let the rays pass through the glass) so no light could get out. However even with the laboratory in darkness, a photosensitive screen near the tube was glowing as it would be if hit by cathode rays. Röntgen investigated what other materials these "X" rays could pass through, including his wife's hand, which became the first X-ray photograph. The question remained: were X rays the only invisible radiation?

68 Radioactivity

THE YEAR AFTER THE DISCOVERY OF X RAYS, A FRENCH PHYSICIST HAD THE IDEA THAT THE PHOSPHORESCENT GLOW OF CERTAIN MATERIALS might be a source of Röntgen's mysterious rays. His theory was proved wrong, but nevertheless led to a whole new field of chemistry.

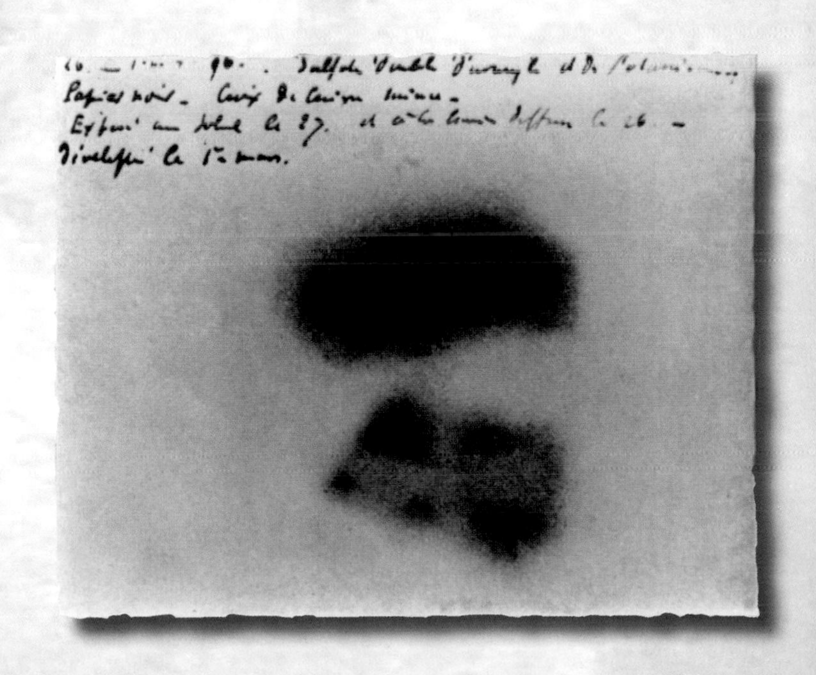

Dark patches on one of Becquerel's photographic plates show where unknown rays have been emitted by uranium minerals.

Henri Becquerel's investigation involved wrapping phosphorescent minerals—those that glow in the dark—in photographic paper. Nothing produced any fogging that would indicate invisible emissions, until he tested pitchblende, an ore of uranium. Further research showed non-glowing uranium minerals did the same. Dubbed "Becquerel rays," this was the first evidence of what was later re-named radioactivity.

69 Discovering Electrons

WHILE LIGHT AND HEAT HAD BEEN REMOVED FROM ALL PUTATIVE LISTS OF ELEMENTS, CHEMISTS WERE STILL TROUBLED by the classification of cathode rays. In many respects these rays behaved as "radiant matter," tiny particles that shared properties with metals and gases. In 1897, an English physics professor succeeding in weighing the rays, with surprising results.

In his day Joseph John Thomson (known by his friends and history as J.J.) was one of the leading physics professors in the world. He lived up to this lofty position by being the scientist who shattered the illusion that atoms are indestructible and indivisible, showing that they are instead merely constructed of yet smaller particles.

Deflecting rays

Thomson's road to discovery began in 1897 when he looked again at an experiment by Heinrich Hertz (now immortalized by a unit of frequency). Cathode rays appeared to be repulsed by the cathode and pulled towards the anode, and so Hertz had tested whether cathode rays could be deflected by another electric field formed by two plates inside the Crookes tube. One plate was positively charged, while the other was negative. Hertz's ray was unaffected by these charges, suggesting that it was

Electrified plates
Anode
Cathode
Deflected cathode ray
Screen

Thompson's cathode ray deflection tube at work at Cambridge University's Cavendish Laboratory, named for Henry Cavendish, the discoverer of hydrogen.

uncharged itself. However, when Thomson repeated the experiment with an improved tube with less gas in it, the ray curved toward the positive plate, indicating that it carried a negative electric charge. (The excess gas left in Hertz's tube was becoming charged by the plates thus negating any effects they might have on the cathode rays passing by.)

George Stoney came up with the word electron *in 1894 to refer to the component of an atom that carried electrical charge.*

Measuring mass

Thomson already knew that the rays responded to a magnetic field, and so he set about comparing the effects of both. That allowed him to calculate the speed and the specific charge of the rays. Specific charge is the ratio of the electrical charge of an object and its mass. The astonishing result was that the "corpuscles" in the rays (as Thomson termed them) were 1,800 times lighter than hydrogen atoms, the lightest of all elements!

In addition, cathode rays passed through solid gold foil and traveled greater distances through air than seemed realistic for something the size of an atom. The only conclusion was that the corpuscles were considerably smaller than atoms. The term *electron* had been coined a few years before as a theoretical charge carrier for electricity. That name found its rightful place with Thomson's discovery of the first "subatomic" particle, an object that was smaller than an atom.

70 Plum Pudding Atoms

THE DISCOVERY OF ELECTRONS LEFT A BIGGER QUESTION: WHERE DID THEY COME FROM? At the turn of the 20th century, the best answer lay in a traditional Christmas dessert.

J.J. Thomson's experiments with cathode rays showed that the mass of electrons did not vary with the material used to make the electrodes in the cathode-ray tube. J.J. Thomson followed the hypothesis that the electric field was pushing the charged particles away from the cathode. Since the cathode was only charged when electrified, the electrons must be a negatively charged component of otherwise neutral atoms. Therefore, as electrons left an atom they must leave the positively charged component behind. Thomson suggested that the negative electrons were diffused through the positive part of the atom like plums in a Christmas pudding. Little more than a guess, the "Plum Pudding Model" was the best science could offer at the time.

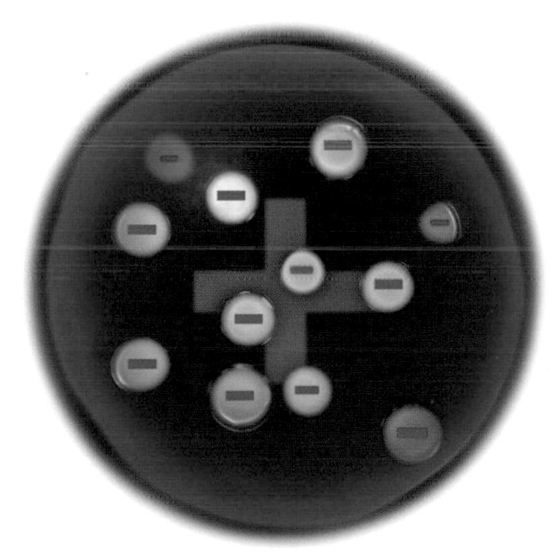

The plum pudding model assumed that the atom was a solid unit of matter.

71 The Noble Gases

When chemists began to isolate new gaseous elements in the 1890s, it became apparent that there was no place for them on the periodic table. The gases also appeared to play no active part in chemistry, leading them to be dubbed as too "noble" to mix with the common elements.

Helium had been identified as a new element before Mendeleev had put his first table together, but only as a tell-tale spectral line in sunlight. No one had isolated any of this new material, let alone calculated its atomic weight or valence. Consequently Mendeleev ignored it. The first sample of helium was collected by Englishman William Ramsay from a uranium-rich mineral he had received from Norway in 1895. (The gas was being produced inside by radioactive decay.) Apart from being lighter, helium had almost identical properties to argon, another new gas isolated only the year before.

WHY SO INERT?

Atoms react with each other and form bonds because their atomic structures are incomplete in some way. Metals have too many electrons, while nonmetals have too few. Reacting is a way of bringing about balance. Noble gases fill a unique place in the periodic table because their atoms, such as helium (right), are naturally balanced and have no need to react with other elements.

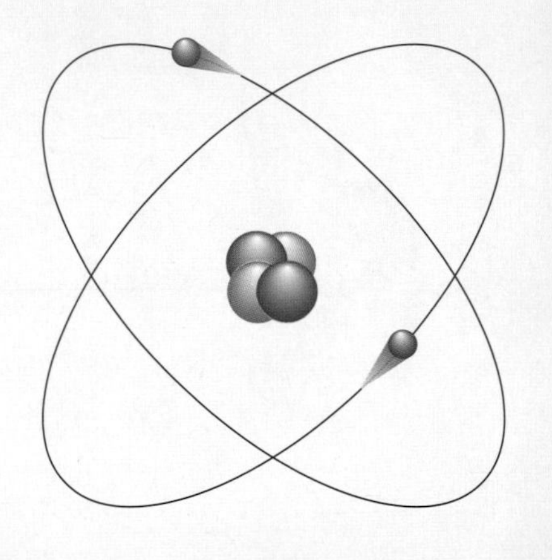

Hidden component

Ramsay had been instrumental in discovering argon along with fellow Englishman John Strutt (generally styled Lord Rayleigh). Rayleigh

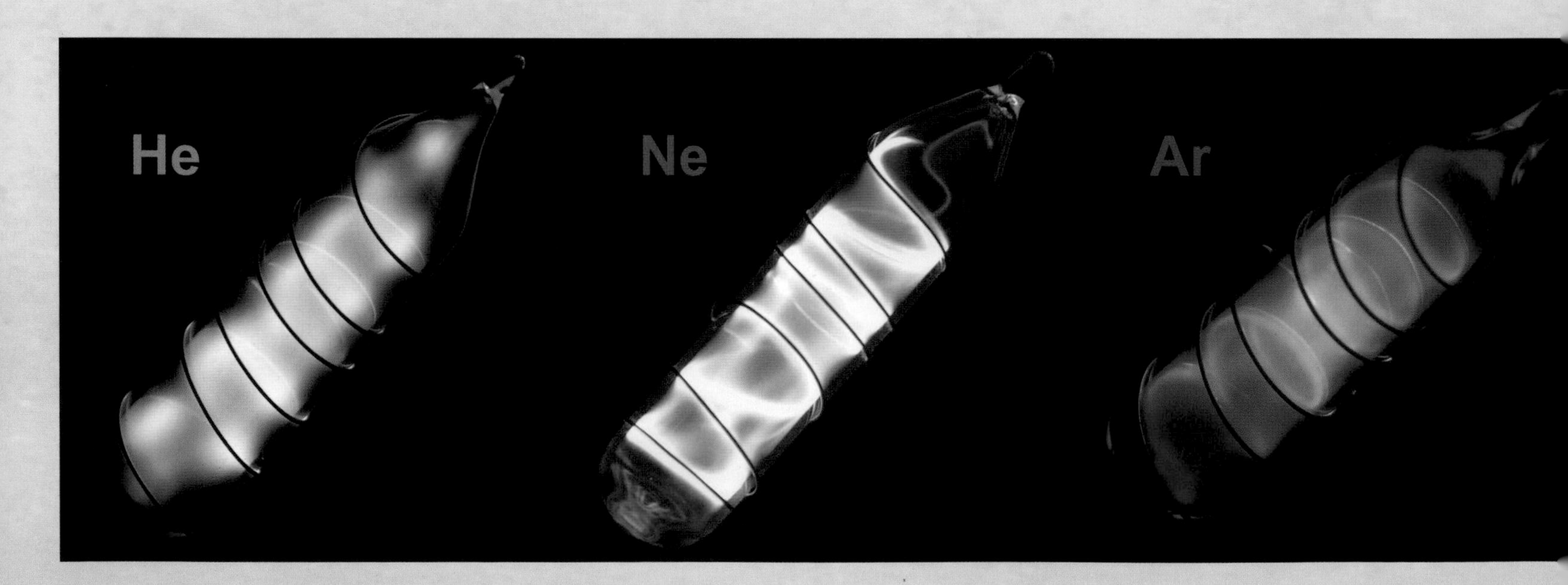

had noticed a discrepancy between the density of nitrogen gas isolated from the atmosphere and samples derived from chemical reactions. Mendeleev suggested the tiny differences were due to nitrogen atoms occasionally forming N_3 molecules instead of N_2. Ramsay and Rayleigh investigated further. They systematically removed all the known gases in air, leaving a tiny sample of an unknown gas. The pair named it *argon*, from the Greek for "lazy," since the gas did not react with other elements at all. In other words it had a valence of zero.

An argon laser is used to measure the speed at which a jet of steam is escaping. Helium, neon, argon, and krypton are all used in lasers to produce specific wavelengths of light.

Chilling the air

In 1898 Ramsay discovered three more of these noble gases by liquefying air with the latest refrigeration technology. He then allowed the liquid air to heat up, collecting each gas as it boiled away in turn. Argon makes up 0.9 percent of the atmosphere. The new gases were even rarer. They were discovered in increasing atomic weight: neon (the new one), krypton (the hidden one), and xenon (the strange one).

All these gases were chemically inert, or noble in more traditional parlance. (Other noble elements include gold and platinum since they rarely react with other elements.) The atomic weights of the noble gases indicated that they form an eighth group in the periodic table, making the final entry in every period. In 1902, Mendeleev relented and allowed the addition of Group 0 for zero valence elements. (In later versions it is often shown as Group 8 or 18.) Another member of this group was predicted by Ramsay, and it would soon come to light during the investigation of radioactivity.

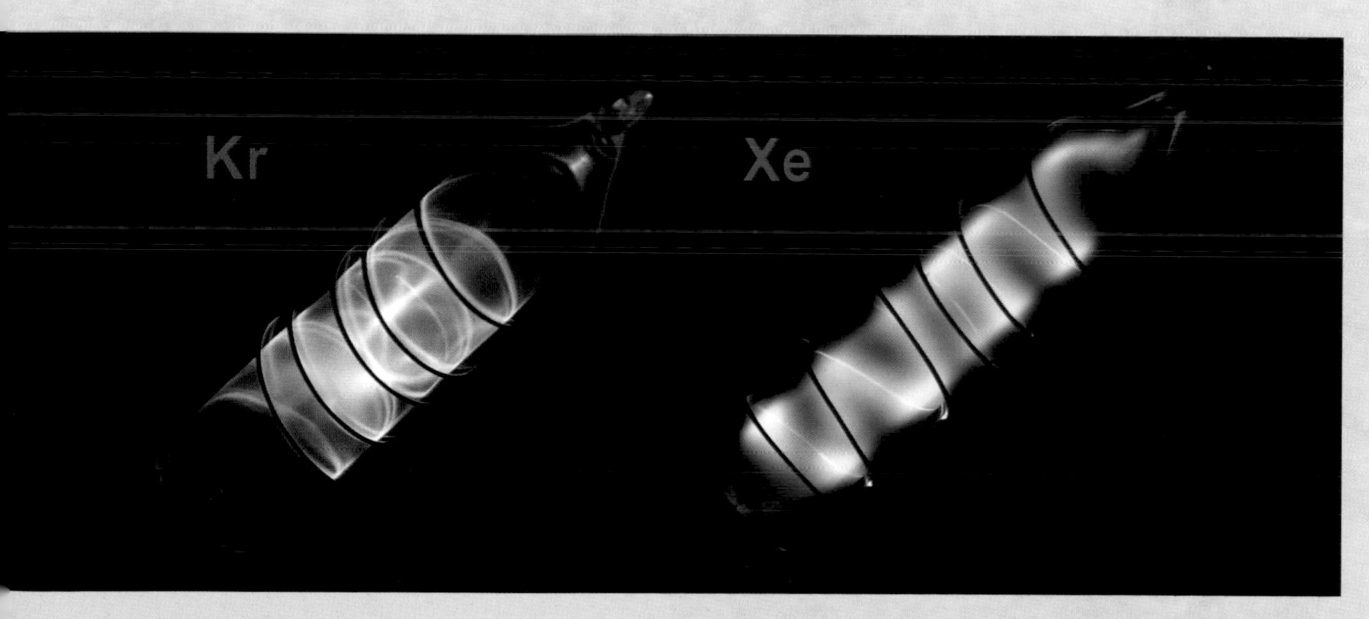

The noble gases glow with distinctive colours in gas-discharge tubes. Neon's red hue is especially distinctive, and it found a use in the early 20th century in the "neon lights" that illuminated the high spots of newly 24-hour cities, such as New York and Paris

72 The Curies

IF THERE IS ONE NAME MOST ASSOCIATED WITH RADIOACTIVITY IT IS CURIE. It took this husband and wife team years of toil and personal hardship to track down the sources of "Becquerel rays."

The Curies were never wealthy despite becoming world famous, with Pierre killed in a traffic accident soon after achieving recognition from the scientific community.

Marie Skłodowska was a brilliant yet struggling Polish scientist, in self exile in Paris from her native Warsaw. In 1894, looking for some space to do her own research Marie approached Pierre Curie, a physicist with a promising track record in the field of electromagnetism. The pair married the following year thus forming a Nobel Prize-winning partnership as a well as personal one.

Soon after having their first child in 1897 (Iréne, herself to be a Nobel laureate) Marie began work on the rays recently discovered by Becquerel. She devoted herself to finding more minerals that produced the same invisible emissions. In 1898 she found that thorium compounds, as well as uranium ones, were radioactive—a term she coined as a portmanteau of "radiation" and "active." At this point, Pierre suspended his own research and joined Marie in hers.

POLITICAL ELEMENT

Choosing to name their first discovery after Poland was a highly political act at the turn of the 20th century. At that time, the country was divided between the powers of Austria, Prussia, and Russia. Polonium was the Curies' call for freedom for Marie's homeland.

he Curies set up a aboratory in a draughty hed. Marie recorded that in vinter it was barely above reezing inside. Marie used his as an opportunity to see he effect of temperature on adioactive emissions—they ontinued unabated.

Missing pieces

The Curies found that pitchblende, a mineral rich in both uranium and thorium, was emitting more radiation than could be expected from its uranium and thorium content alone. The implication was that there was another radioactive source within, a new element. So began the Herculean task of isolating what was a potent but miniscule amount of substance. This involved the painstaking chemical removal of all unwanted elements on an almost industrial scale. Pierre suffered from rheumatism, and so it was left to Marie to process more than a half a ton of pitchblende. After four years, the pair had enough of their new metal element—already named polonium *in absentia* in 1898—to convince their critics. To their joy, the sample contained yet another heavy and radioactive metal, which they named radium (since it was like a radioactive barium). Pierre was killed in a road accident in 1906. Now alone, Marie's later career was blocked by the male establishment, but her legacy was secure.

The Curies spent years isolating tiny traces of radioactive metal from tons of minerals.

73 Transmutation of Matter

RESEARCH INTO RADIOACTIVITY HAD A VERY UNEXPECTED RESULT: the age-old quest of the alchemist was achieved.

Ernest Rutherford oversees some experiments at the Cavendish Laboratory at Cambridge University, the place where he made several of his discoveries. In 1917 he demonstrated nuclear transmutation by bombarding nitrogen atoms with alpha particles to transform them into oxygen atoms.

Radioactivity provided scientists with a window into the structure of matter. A New Zealander, Ernest Rutherford, was one of the first to look through that window. In another age, what he saw might have led to him being branded a wizard, but scientific rigor revealed a surprising truth: elements could transmute from one form to another. The alchemists had been right all along!

The materials characterized by Rutherford as alpha radiation were later shown to be particles with the same structure as a helium nucleus—two protons and two neutrons. Each alpha particle was expelled from a radioactive nucleus. They have a positive charge from the protons and there are no negative electrons to balance that out as would be the case in a helium atom.

Radiation types

The young Rutherford excelled in the universities of his homeland and by the age of 24 he was working for J.J. Thomson in Cambridge. His investigation into uranium produced a breakthrough in 1898. There appeared to be two types of radiation coming from the metal. Alpha radiation (as termed by Rutherford) was blocked by a thin sheet of gold foil, while beta radiation passed straight through it. In 1900 Henri Becquerel showed that beta radiation was composed of the same particles as in cathode rays—in other words, they were electrons. Alpha radiation was presumably made of larger particles that were blocked by the gold foil. (In 1908, Rutherford confirmed this). Also in 1900, another Frenchman, Paul Villiard, found a third type of radiation coming from radium, which had more penetrating power than either of those observed by Rutherford.

Transformations

Rutherford now took a post at Montreal's McGill University, and appointed Briton Frederick Soddy as one of his assistants. In 1901 the pair noticed that thorium gave out a gas as well as radiation, while chemical analysis showed that radium had formed where the thorium had been.

IONIZING RADIATION

Not all radiation is produced by radioactivity. Light and heat are types of radiation as well. However, radioactive radiation carries a lot of energy, enough to ionize atoms (rip off electrons), which can alter chemicals and damage living tissue. Alpha particles do the most damage but are blocked by skin, other radiation goes further but is less damaging.

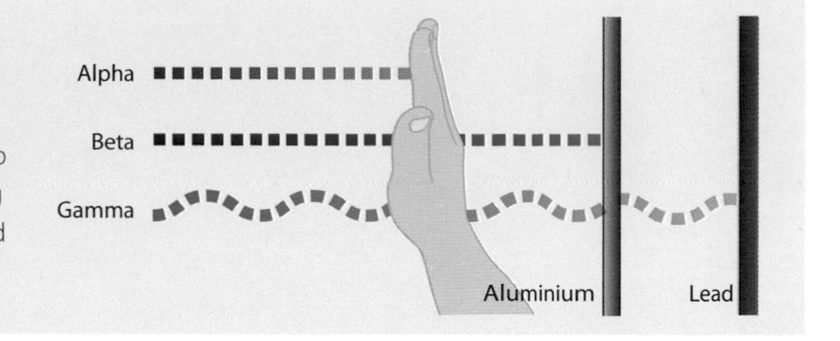

"Rutherford, this is transmutation!" Soddy is reported to have said. "For Christ's sake, Soddy, don't call it transmutation. They'll have our heads off as alchemists," was Rutherford's retort. Research also showed that the amount of radiation fluctuated, sometimes appearing to stop only to start up again stronger than ever. Therefore, in 1903, Rutherford and Soddy put their findings together into the renamed "transformation" theory, which stated that radioactive emissions were the result of atoms of one element decaying into atoms of another. This process continued until stable atoms were formed—uranium decay, for example, runs through 12 unstable elements eventually arriving at lead. Although not quite what Aristotle had predicted, these transformations forced yet another rewrite of the rules of chemistry.

74 The Photoelectric Effect

Although famous for his relativity theory, Einstein won his Nobel Prize for his photoelectric work

IN THE 1880S HEINRICH HERTZ FOUND THAT SHINING BRIGHT LIGHT ON ELECTRODES resulted in increased electrical behavior. In 1905, Albert Einstein used this photoelectric effect to examine the nature of light.

Since the 18th century, light had been viewed as a wave. There was good evidence for this, and Einstein did not seek to disprove it. He just said light was made of particles as well. He called the particles photons, each one carrying a specific amount, or *quantum*, of energy. When light or another stream of photons hits a conductor, the photons' energy is transferred to the electrons, perhaps enough to make them flow as an electric current. Therefore, in reverse, materials emitted energy by radiating out photons.

75 Half-Life

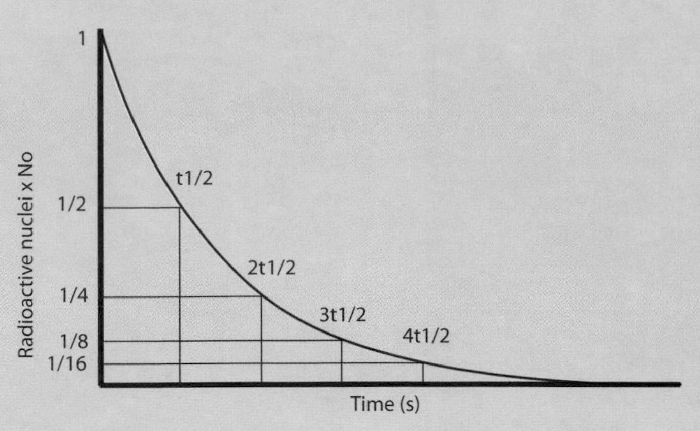

NOT ALL RADIOACTIVE ELEMENTS BREAK DOWN, OR DECAY, AT THE SAME RATE. In 1907, a means of measuring this rate was formulated.

It is impossible to predict exactly when a radioactive atom will decay, so it is expressed in terms of probability. In all likelihood, highly radioactive elements (with a large number of atoms) will decay faster than less potent sources. Ernest Rutherford and others recorded the rates of decay of different radioactive sources and represented each one as a half-life. This was the time in which half of the atoms were likely to have decayed. The half-life of most uranium is 4.46 billion years, the approximate age of our planet. That means that half of the uranium present as Earth formed has now gone.

As this graph shows, half-life remains constant irrespective of how much of the original material has decayed so far.

76 Haber Process

IF WE LOOK FOR CHEMISTRY'S GREATEST CONTRIBUTION TO CIVILIZATION, THE HABER PROCESS WOULD TOP THE LIST. It combines scientific know-how with industrial might to take nitrogen from the air and turn it into fertilizing chemicals, now used to grow enough crops to feed billions of people.

Looking here like a twin of Dr Evil from the Austin Powers movies, Fritz Haber was also the inventor of chemical weapons used in the First World War. While he applied his chemical knowledge for mass killing, the totality of his work undoubtedly saved countless more lives than it destroyed.

The manufacture of fertilizer is a huge chemical industry—as is the organic "fertilizer-free" food sector. However, using fertilizer is nothing new. Preindustrial agrarian society relied on silt-laden flood water to renew the fields with nutrients, burned patches in the forest to create a fertile ash soil, or spread animal dung and "night soil" on crops to increase the harvest.

All life requires a supply of nitrogen as an ingredient for the amino acids that chain together to form body proteins. Animals get their supply by eating plants. Plants absorb water-soluble nitrogen compounds (mainly nitrates) through their roots. These compounds get into the soil by a complex recycling process, partly by bacteria decaying the remains of dead life forms and partly by other bacteria "fixing" nitrogen gas directly from the atmosphere and turning it into plant-friendly forms.

Supply and demand

In 1898, William Crookes, of cathode ray fame, predicted that the human population would soon grow beyond the ability of civilization's ability to produce food. What was

GUANO

Derived from the indigenous Quichua word for dung, guano is essentially dried bird feces. The cool, dry climate of the islands along the Chilean and Peruvian coast meant that seabird guano built up in vast quantities. In the 1860s, wars were fought over these "guano islands," which at the time were the best source of nitrates—used in gunpowder as well as fertilizer.

needed was a means to boost harvests, but the supply of naturally occurring fertilizers–from sewage sludge to guano–was at its limit.

Chemical means

More than three-quarters of the atmosphere is nitrogen, but it is relatively inert chemically and there was no practical chemical technique for fixing it into a water-soluble compound. Then in 1908, Fritz Haber, a professional chemist working in Karlsruhe, Germany, developed a way of reacting nitrogen from the air with hydrogen to produce ammonia (NH_3). The gases had to be at a high pressure, more than 200 times that of atmosphere, and they only reacted when passed over an iron catalyst.

The Haber process was a crucial weapon in World War I. With the British navy blockading natural nitrate sources from South America, the German military machine would have run out of explosives in a matter of months without the chemicals supplied by the Haber process.

The following year the promising Haber process was transformed into the Haber-Bosch process (the current accepted term), when industrial chemist Carl Bosch devised a way to carry out Haber's reaction on a massive scale. The first ammonia plant came on stream in 1911, able to take 33 tons of atmospheric nitrogen and turn it into ammonia in a single day. Haber was awarded the Nobel Prize for Chemistry in 1918, while Carl Bosch's contribution was finally recognised in the 1931 awards.

Ammonia is itself a toxic substance. For use as fertilizer and in other industries (explosives chief among them) the Ostwald process (patented by fellow German Wilhelm Ostwald in 1902) was required. This oxidized ammonia using a platinum catalyst, and then reacted the resultant nitric oxide with water to form nitric acid, an altogether more practical chemical.

Agrochemicals derived from the Haber process transformed agriculture in the 20th century. The Green Revolution between the 1940s and 1970s saw the transfer of this agricultural technology to India and other developing regions in a mostly successful effort to fend off famines.

77 Finding the Nucleus

A LAST-DITCH CHANGE TO A LONG-RUNNING EXPERIMENT SUCCEEDED IN CONSIGNING THE PLUM PUDDING ATOMIC MODEL TO HISTORY. Yet again, Ernest Rutherford was instrumental in this great discovery, along with another whose name is forever linked with radioactivity.

In 1907 Rutherford was lured back to England by a position at Manchester University. The following year he won a Nobel Prize for his work in Canada, but he was already on to new discoveries. With the help of Hans Geiger, a young German, he developed a device for measuring the magnitude of radioactivity—henceforth known as the Geiger counter. Rutherford made use of the counter in isolating some alpha rays. Spectral analysis revealed they were in fact particles with the same properties as helium gas.

Gold-foil experiment

In 1909 Rutherford set up another experiment, and Geiger joined him on this too. Rutherford was going to put his newly found alpha particles to use as a probe to test the plum pudding atomic model. For this model to be correct, the electrons would have to be positioned very precisely in the positively charged "pudding" to ensure that the atom had

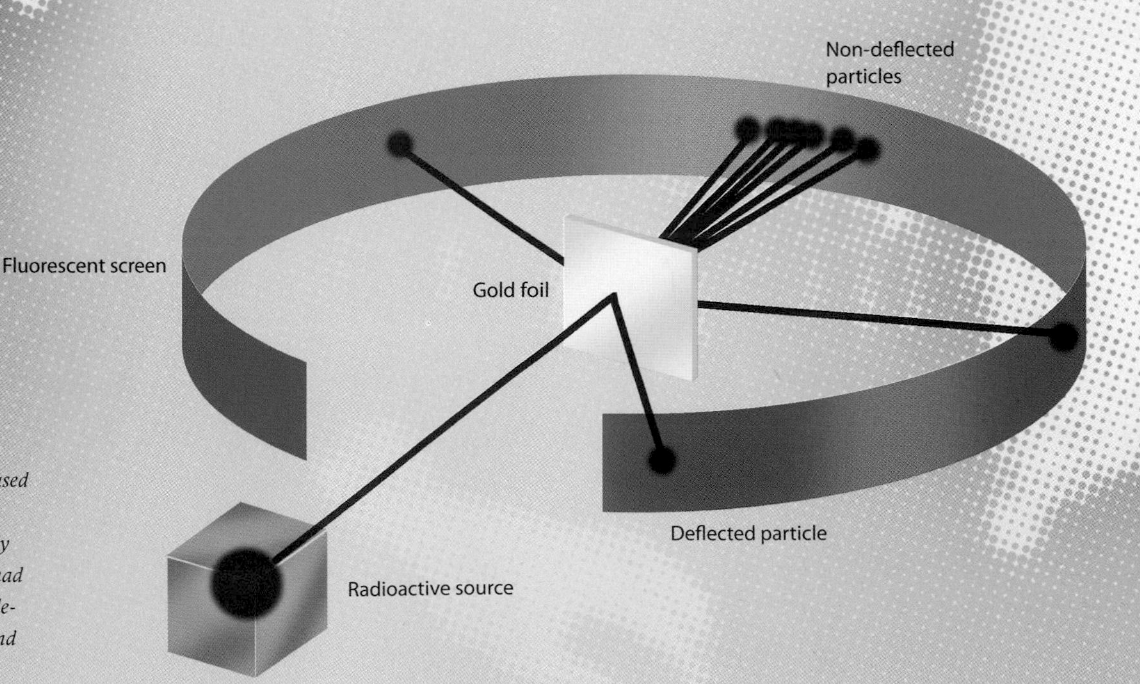

A simplified diagram shows the apparatus used in the Geiger-Marsden experiment. It was only an afterthought that had the researchers place detector screens all around the gold foil.

Ernest Rutherford (right) and Hans Geiger are pictured with the fluorescent screen (above) used in the experiment that revealed the first evidence of the atomic nucleus.

no unevenly charged regions. Rutherford reasoned that if that were true, positively charged alpha particles would pass straight through gold foil, with minimum deviation from the negative charges of the electrons in the atoms. He asked Geiger and a fellow New Zealander, George Marsden, to put this to the test.

Perhaps disappointingly, the great majority of alpha particles sailed through the foil, providing no real evidence against the plum pudding. Rutherford suggested before giving up they should try placing detector screens behind the foil, just to be certain nothing was bouncing back. The repeated experiment did indeed show that a very tiny number of the alpha particles reflected off the foil. Rutherford is reported to have done the *haka*, a Maori war dance when he heard the news. He later described the result thus: "It was almost as incredible as if you fired a 15-inch shell at a piece of tissue paper and it came back and hit you."

Planetary model

By 1911 Rutherford had interpreted the results into an atomic model in which the positive charge formed a minute central core, or nucleus (which had repelled the alpha particles). The electrons orbited around the edge of the atom like planets, held in place by an electromagnetic force. This view of the atom would only last a few years but remains strong in the popular imagination.

An early Geiger counter made from a copper tube. This particular device was used during the discovery of the neutron in 1932.

GEIGER COUNTER

Properly termed a Geiger-Müller tube, this device was inspired by Rutherford, invented by Hans Geiger in 1908, and improved by Walther Müller in 1928. It is based on a sealed tube filled with a low-pressure gas. Two electrodes inside are electrified but no current runs through the gas. When radiation enters the tube, it ionizes the gas, creating pulses of electric charge which can be counted (and used to produce a click sound). The frequency of the pulses is proportional to the amount of radiation in the area.

78 Isotopes and Mass Spectrometry

AS RESEARCHERS TRACKED ATOMS DECAYING FROM ONE ELEMENT TO THE NEXT, THEY FOUND MORE POTENTIALLY NEW SUBSTANCES than there were spaces on the periodic table. Closer examination suggested that the new substances were hitherto unrecorded forms of known substances. However, a new form of analysis would be required to confirm this.

It was Frederick Soddy who realized that radioactive elements were transmuting into too many new elements. He cited the uranium decay series which ends in a stable form of lead; the periodic table places these two elements 11 elements apart but almost 40 intermediate forms had been recorded in the transition from one to the other. Several had been given names by their hopeful discovers, such as metathorium and ionium. However, when the chemical properties of these new substances were analyzed it proved impossible to isolate them: metathorium could not be separated from radium, while ionium displayed all the chemical properties of thorium. In 1912 Soddy's suggested answer to this conundrum was another rule breaker: elements could have more

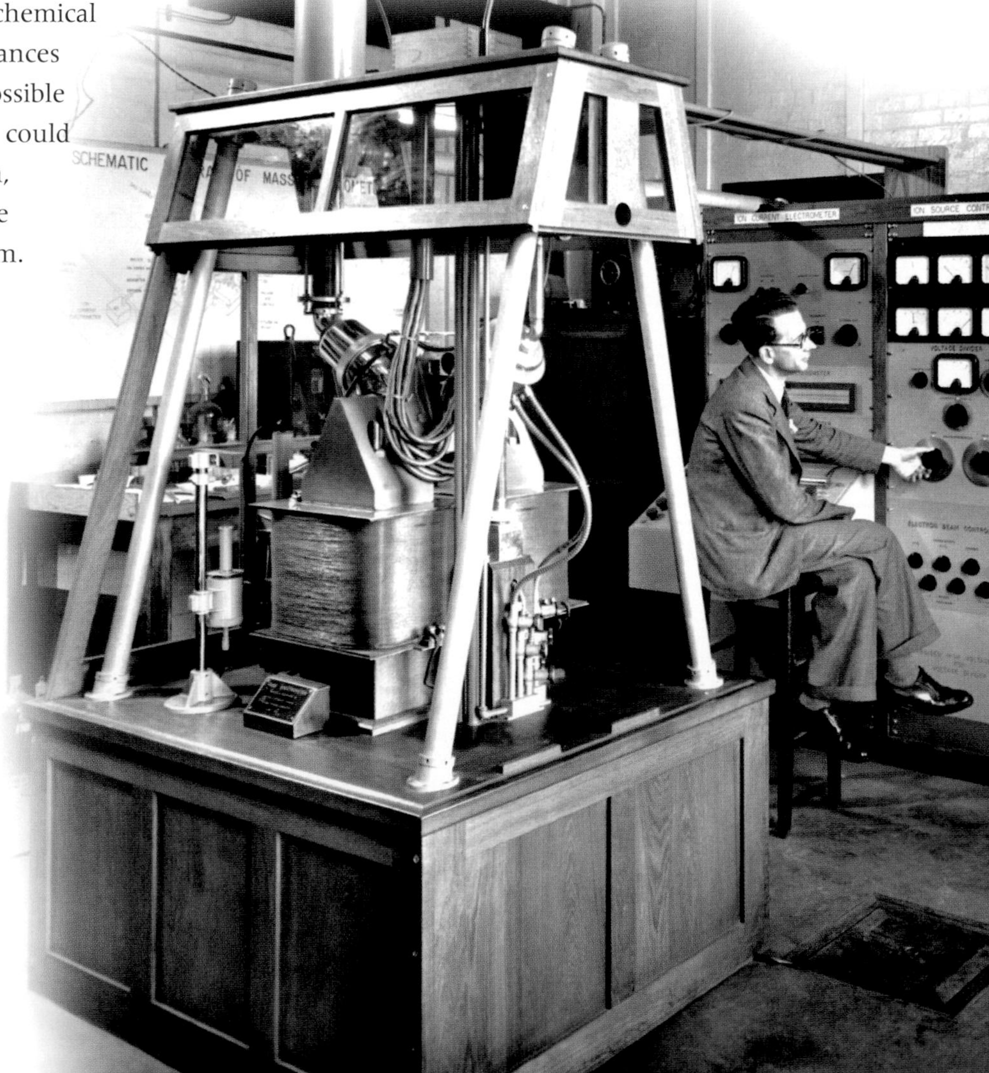

By the 1950s, mass spectrometry had become an everyday analysis technique as this pair of nonchalant researchers at the U.K.'s National Physical Laboratory ably illustrate.

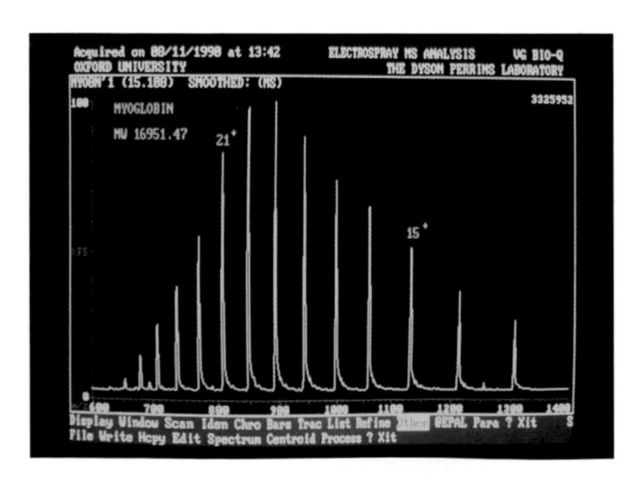

A modern mass spectrometer presents results as a series of peaks showing the mass and quantities of the particles detected.

RADIOCARBON DATING

Carbon 14 is a radioactive isotope produced by cosmic rays bombarding carbon dioxide molecules in the upper atmosphere. Because living things are frequently taking in a supply of carbon, as food for example, their bodies have a small but constant level of C-14. When they die (or are made into an artifact of cloth, bone, or wood) the level of C-14 begins to reduce through radioactive decay. The half life of C-14 is about 5,750 years. Measuring how much lower the C-14 level is in an ancient object, such as this Egyptian mummy, tells us how old it is to within a decade or two.

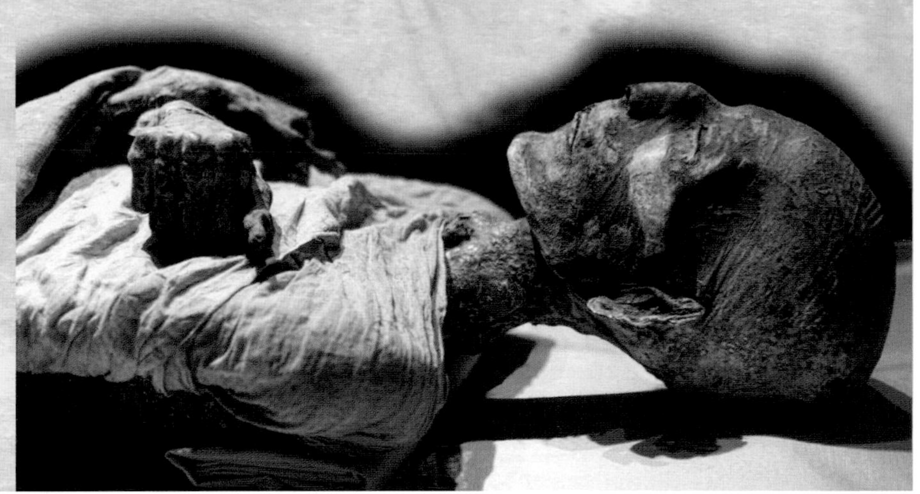

than one atomic mass. A couple of years later, one of Soddy's relatives suggested he should name these different atomic species *isotopes*, meaning "equal place" in reference to their shared location in the periodic table.

Mass movement

The great J.J. Thomson heading up the Cavendish Laboratory at Cambridge provided the first physical evidence of isotopes. He used an anode ray tube, which was a modified Crookes tube that produced a positively charged stream of particles running from the anode to the cathode—the complete opposite of a cathode ray. While a cathode ray is a stream of electrons, an anode ray is made up of positive ions, or atoms that have had some electrons stripped away by an electric field. Magnetic fields did not give a definite deflection to anode rays, which indicated that the ions were of several different masses—they formed from whatever gases were in the tube.

Thomson filled his tube with neon and ran the anode rays through a magnetic field and electric field. The latter would separate ions by their charge—the neon ions were all positive so deflected the same way. The magnetism deflected the light ions further, and the neon gas clouded not one but two areas of a photosensitive detector. This indicated that neon ions (and atoms) had two atomic weights, the isotopes neon 20 and neon 22. Thomson's device was the first mass spectrometer, so named because it separated a gas into a spectrum arranged by charge and mass.

Later mass spectrometers became much more sensitive and are now used to analyze anything from radioactive material to forensic samples. Nevertheless, analysis is not simple because there is no clear indication what the particles are that are revealed in the spectrum. They could be atoms, ionized molecules, or fragments of larger units that had been split apart. Even today a mass spectrometrist must be something of a detective to piece the evidence back together into a real substance.

79 Bohr's Atomic Model

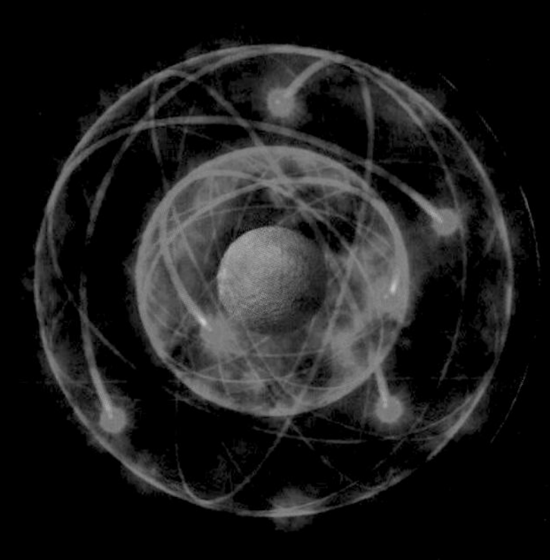

According to quantum physics it is not to know the location and velocity of an electron at the same time. Therefore electrons are best viewed as probability densities, regions encircling the atom where the particles are most likely to be.

THE PLANETARY MODEL OF THE ATOM DID NOT SATISFY THE PIONEERS OF A NEW FIELD OF SCIENCE CALLED QUANTUM PHYSICS. Niels Bohr, a leading figure in this field, applied the rules of the quantum universe to give a new description of atomic structure.

Ernest Rutherford's planetary atom was only two years old when Bohr updated it with the latest understanding of the properties of energy and mass at the smallest scales. The planetary model pictured the negative electrons as spheres of solid matter swinging around the central nucleus. Bohr's version expressed the electron as in a more diffuse region of energy that surrounded the nucleus. He calculated the location of this region using the electromagnetic attraction between the electron and nucleus and the centrifugal force created by its motion around the atom. He found that electrons could only exist at a number of specific locations, or energy levels. The more energy an electron carried, the further out from the nucleus it was located. Bohr checked his calculations with the energy given out by hydrogen atoms during spectral analysis and found they matched. Chemistry had gone quantum.

ELECTRON ORBITALS

Rutherford described the motion of electrons as orbits in keeping with his solar system image. Bohr's model altered this view and so orbits became known as *orbitals*. The lower energy levels produce spherical orbitals, while higher energy electrons have the shapes of dumbbells and doughnuts.

80 The Atomic Number

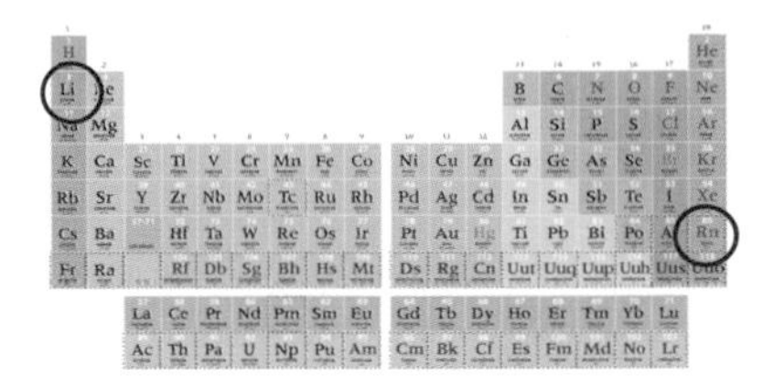

The modern periodic table lists elements by atomic number.

FEW PEOPLE HAVE HEARD OF HENRY MOSELEY. HE DIED YOUNG IN WORLD WAR I, but it was the work of this physicist, at the tender age of 25, that finally made sense of the periodic table.

The Geiger-Marsden experiment had shown that an atom's positive charge was located in a central nucleus. In that experiment, the proportion of the alpha particles that bounced back compared to those that went through was used in statistical analysis to figure out how large the nucleus was. The answer was not very: approximately one 100,000th of the width of the atom as a whole.

In 1913, the same year that Niels Bohr was formulating his atomic model, Moseley was studying the X rays emitted by the atoms of different elements. He found that elements released X rays of a specific wavelength in the same way that they could be identified by the color of their visible light emissions. However, he also found that the X ray wavelength was proportional to the charge of an atom's nucleus.

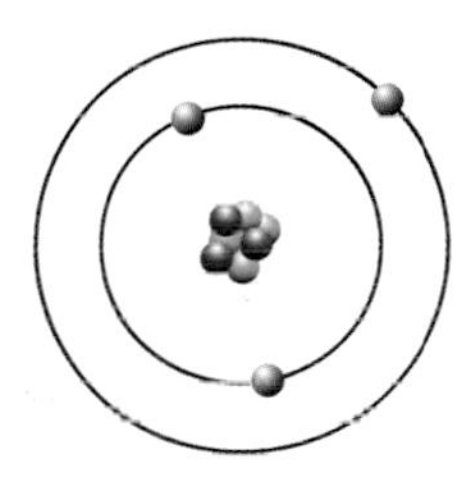

Lithium has three protons in the nucleus, which give the positive charge.

Ordering system

Beginning with hydrogen as one, Moseley allotted an "atomic number" to the other atoms according to the nuclear charges revealed by the X ray spectra. This number system traced the elements almost exactly through the periodic table, in that helium had a charge of two on its nucleus, lithium three and so on.

Until that point the ordering of the periodic table had been based on a nebulous linkage between atomic weights and chemical properties. Moseley's atomic numbers offered a more stringent system and on more than one occasion resulted in elements being reordered (nickel and cobalt for example) or created new empty spaces where no atom with that nuclear charge had yet been identified. In total Moseley predicted the existence of four new elements, including technetium, the first artificially produced element.

Moseley died in the Gallipoli Campaign in 1915, so he never discovered what it was that held the positive charge in an atom's nucleus. That was left to his academic supervisor Ernest Rutherford to reveal after the war.

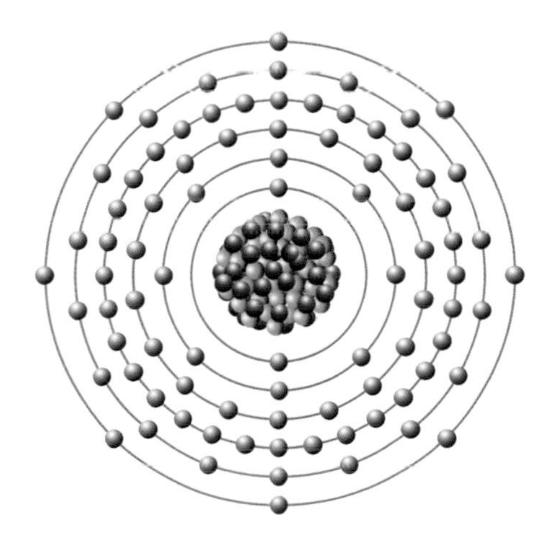

Radon has 86 protons in the nucleus and an equal number of electrons.

81 Quantum Leaps

The world's leading physicists, from Albert Einstein to Marie Curie, gather at the Solvay Conference in 1927 to discuss that year's theme: electrons and photons.

AS THE 1920S APPROACHED, A PICTURE OF THE SUBATOMIC WORLD WAS EMERGING THAT EXPLAINED how light was an electromagnetic phenomenon, intrinsic to the way atoms took in and gave out energy.

GLOWING IN THE DARK

Fluorite is a mineral named for the way it fluoresces, or glows in he dark for a short while. This property is explained by quantum physics. When exposed to sunshine, the atoms in the mineral absorb photons of ultraviolet light, which is invisible to the human eye. A little later, this energy is emitted as visible wavelengths, which is what makes the mineral continue to glow in darkness.

Fluorite gives off an eerie blue light.

Our modern understanding of atoms is still largely rooted in the work of the great scientists of the early 20th century. Putting the breakthroughs of Einstein, Bohr, Moseley, and others together gives us the knowledge that every element has an atom with a specific atomic number, or positive charge, located in the nucleus. The atom as a whole is electrically neutral, which means that the atomic number charge is balanced by a negative charge of equal magnitude supplied by electrons. A single electron has a negative charge of 1 and so the number of electrons in an atom is equal to its atomic number. As described by Bohr, these electrons were arranged in orbitals or electron shells around the nucleus. It had been suggested that the positive charge of the nucleus was also mediated by some kind of particle, but that had yet to be proven.

A so-called Rutherford-Bohr diagram shows how the energy reduction (ΔE) that occurs when an electron leaps from level 3 to level 2 results in the release of a photon of radiation with the frequency f. *Plank's constant (*h*) is a fixed number that relates these two variables. The larger the change in energy level, the higher the frequency (and shorter the wavelength) of radiation released.*

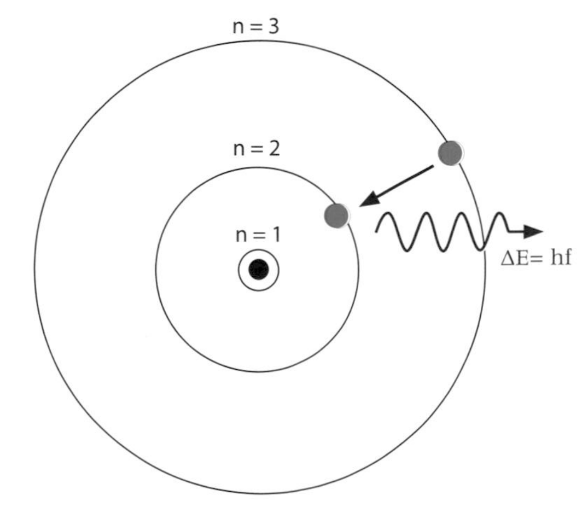

Electromagnetic spectrum

One of the fundamentals of science is that energy can neither be created nor destroyed, only transferred from one mass to another. Quantum physics explains how the structure of the atom allows it to receive and emit energy. This helped with theories on chemical reactions, but also showed why atoms had a specific spectra of light and other so-called electromagnetic radiation as used to identify them in spectroscopy. The electromagnetic spectrum begins with radio waves, which have the longest wavelengths (very short radio waves are called microwaves, but they are still relatively long). Next comes infrared, which we perceive as heat; then visible light; ultraviolet (which causes sun burns and tans); X rays; and finally gamma rays, which have the shortest wavelengths of all. Photons of gamma rays carry the most energy (enough to ionize atoms), while radio waves carry the least.

Energy in and out

The unique structure of each element's atom means it can only absorb specific quanta of energy, which arrive as photons of radiation with a corresponding wavelength. The photon transfers its quantum of energy to an electron, which makes a leap to a higher energy level. As such a quantum leap is all or nothing, it cannot happen in two jumps, nor can the electron jump higher and fall back to its new position. When the electron moves back to its original energy level, it gives out the energy in the form of another photon, with a certain energy and wavelength. This process is what makes hot objects glow with colored light or electrified metals give off a stream of radio waves or X rays. High-energy events such as radioactive decay produce gamma rays. Harnessing this knowledge of energy, atoms, and radiation, chemists set off to find out what it was that made atoms bond together.

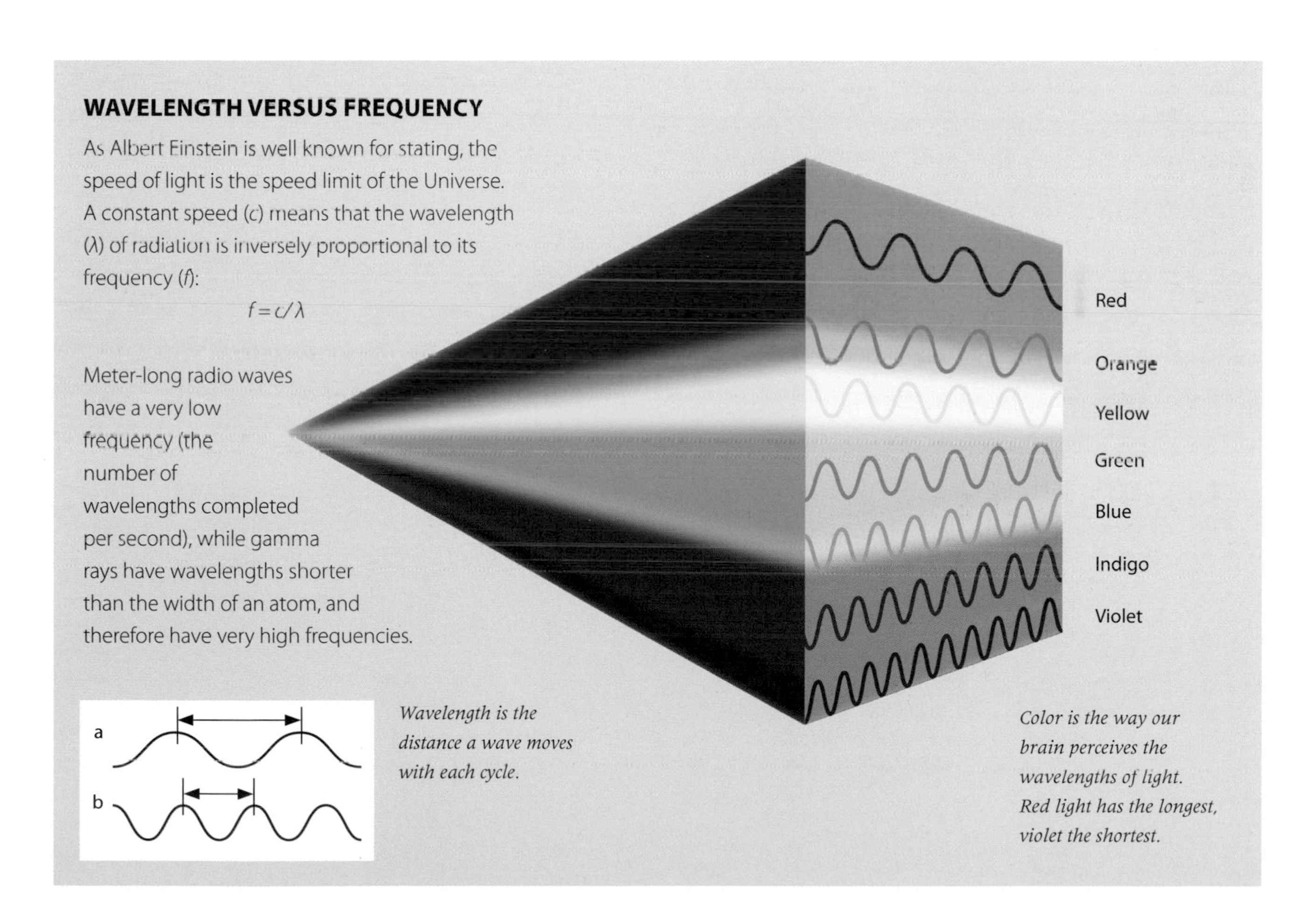

WAVELENGTH VERSUS FREQUENCY

As Albert Einstein is well known for stating, the speed of light is the speed limit of the Universe. A constant speed (c) means that the wavelength (λ) of radiation is inversely proportional to its frequency (f):

$$f = c/\lambda$$

Meter-long radio waves have a very low frequency (the number of wavelengths completed per second), while gamma rays have wavelengths shorter than the width of an atom, and therefore have very high frequencies.

Wavelength is the distance a wave moves with each cycle.

Color is the way our brain perceives the wavelengths of light. Red light has the longest, violet the shortest.

82 Protons Discovered

HAVING BEEN INSTRUMENTAL IN THE DISCOVERY OF THE ATOMIC NUCLEUS, ERNEST RUTHERFORD LED THE CHARGE to find out what it was made of. Then, in 1917, he found hydrogen nuclei coming from other atoms.

Way back in 1815, William Prout suggested that the atoms of all heavier elements were made of multiple clusters of hydrogen atoms, the lightest substance of all. In 1917, Rutherford fired alpha particles at nitrogen atoms (atomic number 7). He found a few turned into oxygen atoms (atomic number 8), releasing a hydrogen nucleus. An alpha particle was a helium nucleus (atomic number 2) and so in converting nitrogen to oxygen, one of its charged units passed to the larger atom, leaving a hydrogen nucleus (atomic number 1). It followed, therefore, that Prout was correct, and the nuclear charge was carried by particles. Hydrogen atoms only had one of these, and Rutherford named them protons, meaning the "first." Perhaps bizarrely, a proton has an equal and opposite charge to an electron but has almost 2,000 times the mass.

83 X-Ray Crystallography

THE SYMMETRY OF CRYSTALS ON THE LARGE SCALE led people to wonder if a similar order existed on the atomic scale. This was answered by firing X rays through them.

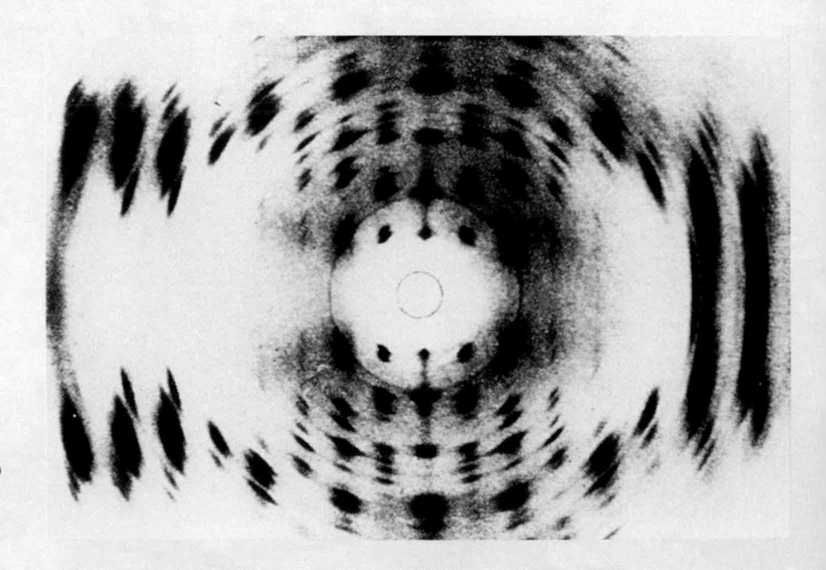

The diffracting X rays form distinctive patterns on photographic plates.

Electromagnetic radiation behaves like a wave as well as a stream of particles. As a wave, it will diffract, or ripple outward as it passes through a gap that is similar in size to its wavelength. In 1912, X rays were found to diffract as they traveled through crystals. That provided a means of measuring the spaces between atoms in the crystal, and the angles at which the X rays emerged could be used to postulate possible structures for the molecules. By 1920, salt crystals had been shown to be cubic, diamond was made of tetrahedra, while graphite was a series of hexagons. Attention then turned to more complex molecules.

84 The Benzene Ring

SINCE ITS DISCOVERY IN 1825 BY MICHAEL FARADAY, CHEMISTS HAD PUT FORWARD SEVERAL POSSIBILITIES FOR the molecular structure of benzene. The true nature of this curious six-carbon compound was to be revealed using X-ray crystallography.

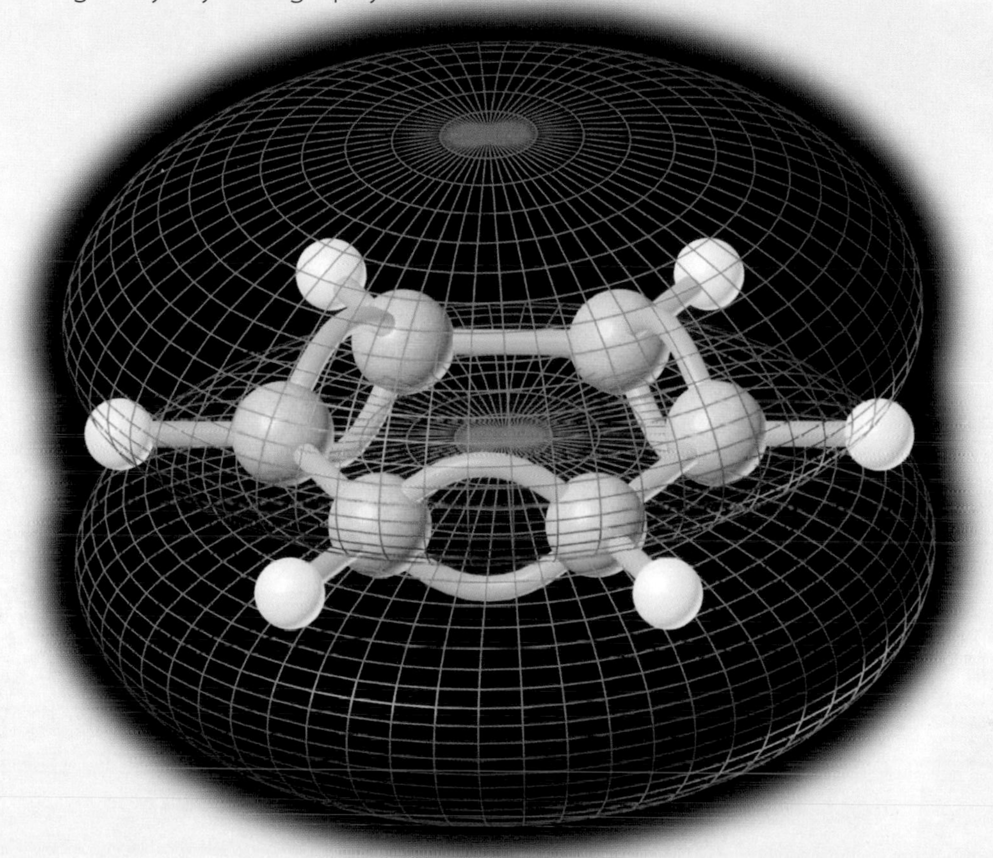

In a benzene ring all the carbon atoms form four bonds. However, the double bonds linking three pairs become shared equally by six atoms resulting in a sandwich of doughnut-shaped charges around the molecule.

Benzene is first and foremost among what chemists call aromatic compounds. Chemists had noticed that this chemical and related substances had a particular set of properties, but it was unclear what underwrote them. Faraday had found that benzene had six carbon atoms but only six hydrogens. That created something of a puzzle: how did all these atoms connect together, considering that each carbon had to make four bonds. It was known that two carbon atoms could connect with double (and even triple) bonds, and chemists suggested everything from complex criss-crossing molecules to trapezoid blocks. Friedrich Kekulé proposed benzene had a ring shape in 1865, and in 1929, Irish researcher Kathleen Lonsdale used X-ray diffraction to prove he was right. As predicted by German Johannes Thiele in 1899, the three double bonds needed to form a ring of carbons were "delocalized" from any one distinct pairing of atoms. Instead, the shared electrons were spread over the whole ring. That created a cloud-like bond that gave the structure the stability that characterizes aromatic compounds, many of which are involved in polymers such as DNA and plastic.

85 Chemical Bonds

IN THE LATE 1920S, LINUS PAULING BEGAN TO SET OUT HIS VISION OF ATOMIC BONDS, WHICH MADE HIM THE AUTHORITY on the subject for years to come. He drew together many strands of chemistry and physics to describe the role of electrons in connecting atoms.

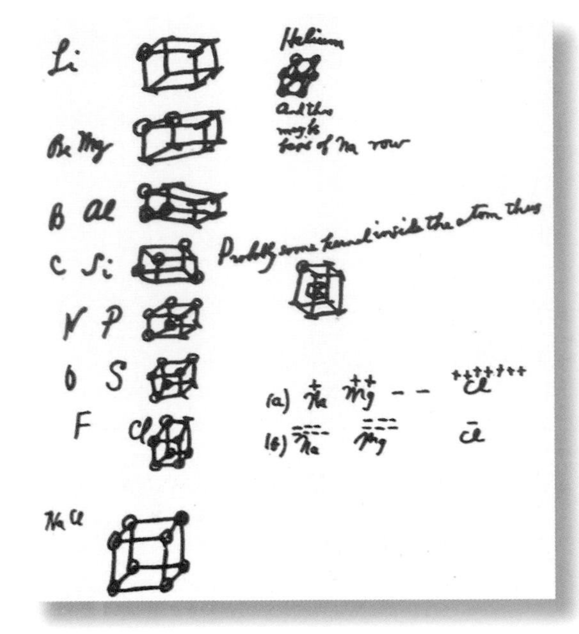

Gilbert Lewis's sketch from 1902 showed the electrons of different atoms arranged on the corners of a theoretical cube.

An atom composed of protons, neutrons, and electrons as described by Rutherford, Bohr, and many others worked well with the understanding of ionic bonds as it stood in the early 20th century. Building on the work of Faraday and Berzelius, Swedish chemist Svante Arrenhius had proposed that the ions in electrolytes, the liquids that conducted electricity, formed when the molecules or crystals of a solid dissociated into their charged constituents (for example when they dissolved or melted into a liquid). The implication of this was that solid materials were held together by the electromagnetic attraction between ions of opposite charges.

Filling shells

According to the Bohr atomic model, an atom becomes a positive ion when it receives enough energy for an electron to leave the atom, creating an imbalance with the unchanged positive charge on the nucleus. By contrast, negatively charged ions form when atoms take in extra electrons. However, not all compounds can be explained through *ionic* bonds. What about the substances that do not dissociate in the way Arrenhius described?

An answer came from an American, Gilbert Lewis. As far back as 1902, Lewis pondered periodicity: the first two elements, hydrogen and helium, had opposite properties, with hydrogen being reactive and helium inert. The second period had a similar pattern, only there were eight elements between reactive lithium and inert neon. This was repeated in the third period, too. Lewis interpreted atomic bonds as atoms achieving stability by having sets of eight

Linus Pauling's work on covalent bonds allowed him to calculate the length and angle of each bond to build an accurate three-dimensional model of molecular structures.

electrons. Thus a sodium atom lost its single outer electron to form a positive ion, while chlorine gained one to form a negative ion. Inert neon did not lose or gain anything since it already had the set of eight.

Covalence

In 1916, Lewis suggested that the same thing was happening in non-ionic compounds, only the outer electrons were shared in pairs, rather than exchanged. This kind of bonding became known as *covalence*. A decade later Linus Pauling did the math to relate this concept to Bohr's quantum atom. To Pauling a covalent bond was an overlap between the orbitals of two atoms. He also showed that since some atoms pulled on electrons more than others, every chemical bond had a specific length, with shorter bonds being strongest. Finally he used the shapes of orbitals to give a 3D structure of even the most complex molecules.

DOUBLE PRIZE WINNER

Linus Pauling is one of four people to win two Nobel prizes. The first in 1954 was for his work on chemical bonds. Then, in 1962 Pauling won the Nobel Peace Prize for publicizing the grave threat of nuclear proliferation. He was credited with helping to sway public opinion and secure the first Cold War test-ban treaties.

86 The Neutron: The Missing Piece

IF THE ATOMIC NUMBER COUNTED THE PROTONS IN AN ATOM, was there another particle that added up to the atomic mass?

In 1920 the great Ernest Rutherford predicted a neutrally charged nuclear particle, which along with protons gives atoms their mass. In the early 1930s researchers found a new type of radiation that was produced when alpha particles hit beryllium and boron. This radiation had no charge but was too powerful to be gamma rays. James Chadwick directed the radiation into a variety of gases made of molecules of different masses. He measured the displacement of the gases to calculate the mass of the particles in the radiation, and found it was about the same as that of protons. Here was the neutron—a neutral particle that completed the atomic nucleus.

James Chadwick's notes show his calculations for the mass of a neutron, which turned out to be slightly more than that of a proton.

87 Practical Polymers

TWENTIETH-CENTURY CHEMISTS WERE NOT ONLY ABLE TO ANALYZE NATURAL SUBSTANCES, THEY COULD ALSO COPY THEM. Although it was often a hit-and-miss business, artificial materials, such as nylon and polyethylene, have revolutionized society.

Another chemistry term we have Jöns Jacob Berzelius to thank for is *polymer*. A polymer is a large chain molecule that contains several—perhaps millions—of repeating units, or monomers, that form a long chain. Many of nature's most familiar substances are polymers, such as the cellulose in wood, the starch in food, and the proteins in muscles.

Synthetic materials

Natural rubber is a polymer of the organic compound isoprene, which is found in the latex fluid tapped from rubber trees. This material had been in common usage since the 18th century. Charles Mackintosh waterproofed his coats with it in the 1820s, and the following decade Charles Goodyear developed vulcanization which toughened the rubber with crosslinks between the polymers, making it suitable for applications such as tires.

Organic chemists experimented with artificial polymers in the mid 19th century. Polyvinylchloride, or PVC, first made in 1835, proved to be brittle and uninteresting. Then in the 1920s, U.S. chemists looking for a new glue found they could make PVC

NON-STICK SLIP

Polytetrafluoroethylene might not be a household name, but that is because it is normally shortened to Teflon. This polymer was made by accident in 1938 when U.S. researchers were trying to make an inert gas for use as a refrigerant—one of the now banned CFCs. Catalyzed by its iron container, the gas polymerized into a slippery wax. Teflon is what makes nonstick pans nonstick—the polymer is too slippery for even charring food to cling to.

A view of the nylon fibers in a pantyhose magnified 300 times through an electron microscope.

more pliable with additives. At the same time, British researchers studying ethylene suffered frequent explosions. One rebuilt apparatus leaked, and the oxygen that was accidentally included acted as a catalyst, resulting in the polymerization of ethylene to create polyethylene. This material is now the most widely used plastic of all, found in bags, bottles, and bowling balls.

STYROFOAM

This lightweight material is a polymer of styrene, an oily compound similar to benzene. Polystyrene is tough but brittle; CD jewel cases are made from it. In 1959 styrofoam was invented in a process which traps air bubbles in the polymer as it solidifies.

Artificial silk

It is perhaps hard to imagine now, but nylon was originally seen as the perfect upgrade from silk. It has a lower opinion today, but the artificial fabric revolutionized clothing. Nylon was developed in 1935 by a U.S. team led by Wallace Carothers. They found that liquid polyesters formed fine silky strands but they were very weak. So they tested the strength of polyamides by pulling a strand from the liquid and running with it down a corridor. The strands stayed intact, and nylon became the first successful artificial fabric.

88 The First Artificial Element

MENDELEEV LEFT A SPACE IN THE PERIODIC TABLE FOR ELEMENT 43. But, try as they might, no one could find it. In the end it was made by the world's first particle accelerators.

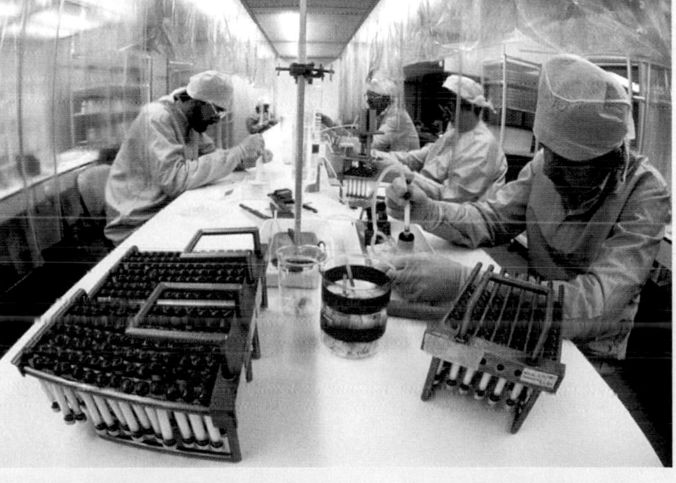

Technetium 99 has a half-life of just a few hours and so it is used as a short-term radioactive source for imaging the internal organs. Here technicians are harvesting Tc-99 from radioactive molybdenum for use in hospitals.

There were many false reports of element 43. In 1877, Russian Serge Kern said he found it in a platinum ore—naming it davyum. In 1908, Masataka Ogawa thought he had it, giving it the name nipponium. Instead, it turned out he had found the first evidence of rhenium.

In the end science itself made this elusive element. In 1930s' California, Ernest Lawrence was developing cyclotrons, the world's first particle accelerators. In 1936 some physicists working in Sicily requested a sample of cyclotron components that had become radioactive for testing. Somewhat alarmingly by today's standards, Lawrence sent it by mail. At the University of Palermo Carlo Perrier and Emilio Segrè found two isotopes of element 43 in the sample. Because it had been made by machine, the element was named technetium. Miniscule quantities of technetium have since been found in nature, but this element is still one that is manufactured rather than refined.

89 The Citric Acid Cycle: Life as Chemistry

EVEN AT THE DAWN OF MODERN CHEMISTRY IN THE 17TH CENTURY, THE ROLE OF CHEMICAL ACTIVITY IN LIFE WAS BEING EXPLORED. Oxygen, water, and carbon dioxide were all identified as being involved, but few predicted the sheer complexity revealed by a breed of "biochemists" in the 20th century.

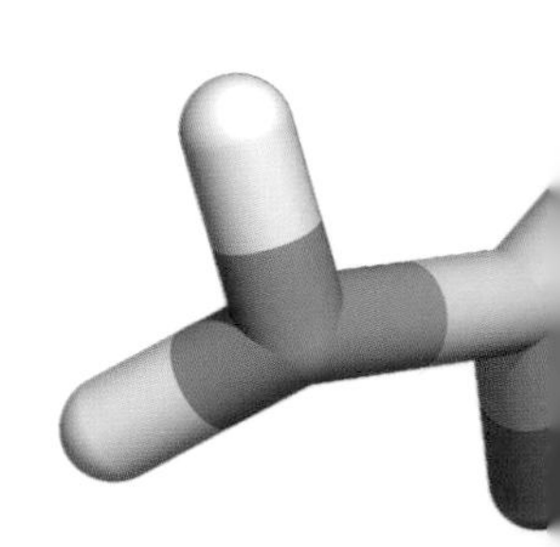

In the early 17th century, the Belgian physician and alchemist Johannes van Helmont recorded the weight of a plant as it grew in a tub of soil. The soil's weight remained more or less constant, while that of the plant increased steadily. The only thing Van Helmont added was water, so he reasoned that this was what powered plant growth.

Making fuel

Later research showed that carbon dioxide, a gas that van Helmont was one of the first to record (calling it *spiritis sylvestre*), was also involved. Joseph Priestley discovered that while animals increased the amount of carbon dioxide or "fixed air" in the atmosphere, plants reduced it. In 1778, Austrian royal physician Jan Ingenhousz showed that a mouse suffocating in its own carbon dioxide in a sealed container could be revived by placing a plant in with it, and it came around all the faster if the plant was in bright sunlight. In 1796, Jean Senebier put this altogether: green plants take in carbon dioxide and release oxygen under the influence of light. Water was later included in the theory, and it was found that plants produce glucose (a sugar) by reacting water with carbon dioxide. This reaction was powered by light energy and was termed *photosynthesis* (meaning making with light).

Cycle of energy

All life is powered by fuels such as glucose, either obtaining it through photosynthesis as do plants, or by eating other organisms, as animals do. The fuel is oxidized in a reverse process to photosynthesis called respiration. At its simplest, respiration converts glucose back into carbon dioxide and water, effectively burning the sugar to release the energy stored in it by photosynthesis. All life respires, producing carbon dioxide. To an animal this is a toxic waste product and is exhaled—as several pioneering pneumatic chemists had revealed. Plants also give out carbon dioxide when it is dark. However, when it is sunny, the plant recycles the gas for photosynthesis, this time giving out oxygen as a waste product—and so ensuring that the air is resupplied with this gas so vital for all life.

Obviously the glucose does not burn in the physical sense. Instead respiration extracts its energy in several steps. In 1937, a team led by Hans Krebs, a German-

Jewish émigré working in Sheffield, England, traced the metabolic pathway the glucose took as it was respired. Krebs found a 12-stage cycle, which began with the six-carbon glucose being transformed into a citrate, another six-carbon compound, related to citric acid. As compounds were converted from one to the next, energy is released and carbon dioxide molecules liberated in the process. The final stage in the so-called citric acid cycle is a four-carbon compound, known as oxaloacetate, which is used to create another citrate—and so the cycle repeats.

Krebs found it hard to publish his findings, but eventually the scientific community realized he had revealed the central process of all life. He won a Nobel Prize in 1953, and for a while the process was known as the Krebs Cycle. Arsenic, cyanide, and rat poison all work by interfering with the citric acid cycle. Without it life cannot continue.

ENERGY CARRIER

Adenosine triphosphate molecules, ATP for short (above), are the power packs for a cell's machinery. As its name suggests, each one carries three phosphates. By dropping a phosphate, the ATP transfers its store of energy to another metabolic process—becoming an ADP (diphosphate) in the process. The citric acid cycle then takes the energy from glucose and uses it to add phosphates to convert ADP back into ATP. One glucose molecule alone can charge up 38 ATPs.

Hans Krebs found each compound in the citric acid cycle by blocking the pathway at each successive stage and identifying the chemical that remained. Much of the work was done using minced pigeon muscles.

ENERGY TRANSFORMATIONS
IN LIVING MATTER

A SURVEY

BY

H. A. KREBS AND H. L. KORNBERG

MEDICAL RESEARCH COUNCIL
UNIT FOR RESEARCH IN CELL METABOLISM,
DEPARTMENT OF BIOCHEMISTRY
UNIVERSITY OF OXFORD

WITH AN APPENDIX BY
K. BURTON

WITH 21 FIGURES

SPRINGER-VERLAG
BERLIN · GÖTTINGEN · HEIDELBERG
1957

Details of Krebs' work were published in this booklet in 1957 following his Nobel prize win.

90 Splitting Atoms

ALBERT EINSTEIN'S FAMOUS EQUATION $E=MC^2$ SHOWS THAT ENERGY AND MASS ARE INTERCHANGEABLE. The letter c stands for the speed of light, a very large number. If there is one thing we can all understand from the equation it is that even tiny masses contain a huge amount of energy.

It all began with the discovery of the neutron. Enrico Fermi, an Italian physicist, saw the neutron as a powerful probe for investigating nuclear physics. Unlike the alpha particle, which had been used previously, the neutron had no charge and so was not swayed by the electromagnetic forces at work inside atoms. Fermi's team set about bombarding all kinds of atoms, and in 1934 they announced that an element with an atomic number of 94 (Fermi named it hesperium) had been formed from uranium. No one, Fermi among them, was quite sure how this was possible.

Otto Hahn, a researcher in Berlin, began to perform similar experiments. In 1938, he found barium in a sample of uranium after it had

The fission reaction is most likely to proceed with slow neutrons, which have less energy than most free neutrons but are more likely to hit a nucleus.

Artwork showing the moment when the world's first nuclear reactor, Chicago Pile-1, became self-sustaining at 15:22 on December 2nd 1942, in a racquets court at the University of Chicago.

been bombarded with neutrons. His colleague Lise Meitner showed that the barium had formed because a neutron had joined with a uranium nucleus, making it so unstable that instead of decaying normally it had split in two—here was nuclear fission.

Alert to danger

The calculations showed that, as Einstein's equation predicted, fission reactions released an inordinate amount of energy, potential power that could perhaps be used. Léo Szilard, a young Hungarian scientist, realized that if a fission released two or more neutrons as the nucleus split, then a chain reaction would result, with each fission resulting in at least two more until all the fissile nuclei had been used up. Uncontrolled, this could lead to a terrible explosion.

Szilard convinced Fermi (now in New York to escape the Nazis) to keep quiet about this development. But Parisian Frédéric Joliot-Curie (Marie's son-in-law) did not. In 1939, he reported that a nucleus of the rare isotope uranium 235 would produce at least three neutrons during a fission. With the world descending into war, the race was on to learn to control fission chain reactions, and so perhaps create the decisive weapon.

Pile 1

In 1942, Enrico Fermi built the world's first nuclear reactor, known as Chicago Pile 1, at the city's university. It used blocks of graphite to focus neutrons onto the uranium fuel and so create a chain reaction. In modern parlance, the nuclear fuel was unenriched, which meant it contained only 0.7 percent of U-235, the fissile isotope.

Nuclear fission was now a new force of nature brought under the control of man. The Manhattan Project followed Fermi's breakthrough, and focused on isolating U-235 for use in the most potent weapons the world had ever seen. (A similar project was developed by Nazi Germany.) Within a decade of Hiroshima and Nagasaki, enriched nuclear fuels were being used as heat sources in civilian power plants, and the first nuclear-powered submarine had enough power to stay underwater for four months.

The next goal is harnessing nuclear fusion. The world's scientists are collaborating on such a project at the ITER facility in France. Only time will tell.

NUCLEAR FUSION

Stars (iincluding our Sun) are powered not by fission but fusion. This process takes small nuclei, such as those of hydrogen, and squeezes them together so they form a single, heavier nucleus. The forces required are enormous, such as you might expect at the center of the Sun, which has a pressure 250 billion times higher than the atmosphere on Earth.

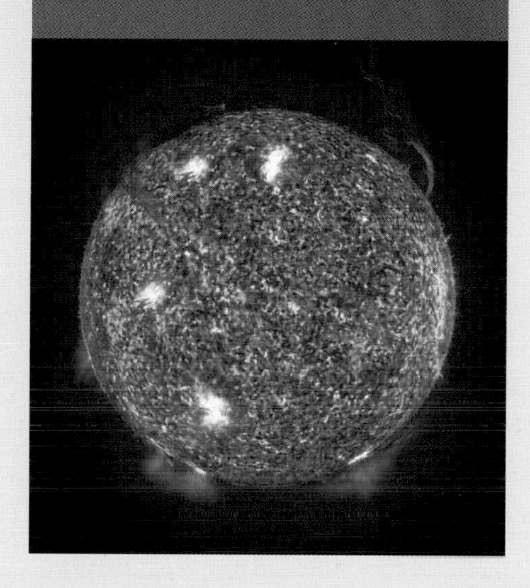

NUCLEAR POWER

A nuclear power station works in the same way as any thermal power plant, using steam to spin turbines linked to an electricity generator in order to produce electric power. The steam is produced with the heat from the fission of nuclear fuel in the reactor. The fission is controlled by boron rods which soak up neutrons. This prevents the chain-reaction from occurring so fast that it would create an explosion.

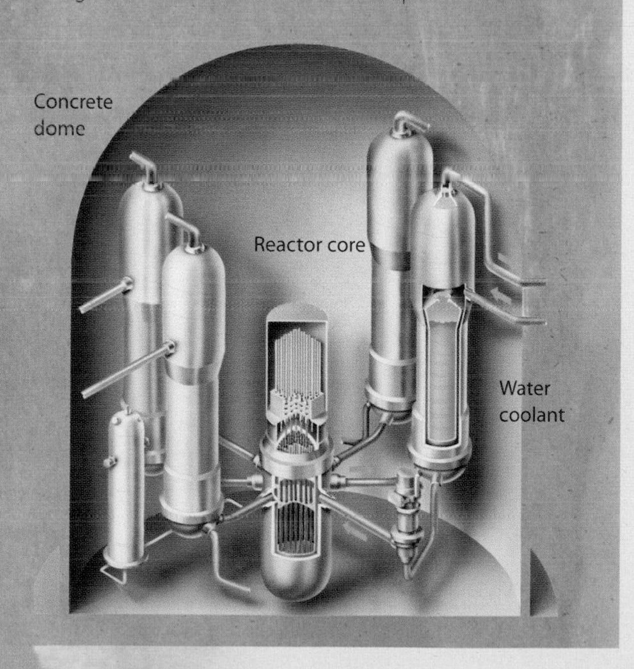

Background: In the atomic bomb attack on Nagasaki, Japan in 1945, just one gram of nuclear material was converted into its equivalent energy. The resulting explosion killed 70,000 people.

91 The Transuranic Elements

At the dawn of the atomic age, the work of Enrico Fermi provided hints that the periodic table could be extended. Uranium, the heaviest naturally occurring element, was not the largest possible atom. Larger, heavier, and relatively stable atoms could be made in laboratories.

In the early 1930s, Fermi's research team in Rome were bombarding all kinds of materials with neutrons to see if they resulted in any nuclear changes. When he reported that a sample of bombarded uranium contained traces of an element with an atomic number of 94, few people took it seriously. (It was eventually revealed to be a mistake.) At the time it was thought that uranium, with its atomic number of 92, was the largest atom in the universe. If more massive elements were possible, why had none been found in nature?

Nevertheless a mechanism was known by which uranium atoms could increase in atomic number. While alpha decay is simply the release of two protons and neutrons from a radioactive nucleus, beta decay is slightly more complicated, involving a neutron breaking down into a proton and an

Glenn Seaborg is pictured in 1950 with an ion exchange column for isolating heavy metals. He discovered 10 new elements: plutonium, americium, curium, berkelium, californium, einsteinium, fermium, mendelevium, nobelium, and seaborgium.

DOMESTIC RADIOACTIVITY

The most radioactive object in a home—although completely safe—is the smoke detector. It is fitted with a tiny supply of americium 241—less than a millionth of a gram in all. This isotope, first manufactured in 1944 by Glenn Seaborg's team at Berkeley, has a half-life of about 420 years. As it decays, the radiation ionizes the air inside the detector, allowing a small electric current to run through it. If smoke particles enter the detector, they block the flow of the current, causing the alarm to go off.

electron. The electron is released—the beta particle—but the proton stays in the nucleus, increasing the atomic number by one. In 1940, U.S. nuclear chemist Edwin MacMillan showed that bombarding uranium 238 (U-238), the main isotope of uranium, resulted in the formation of U-239, a short-lived isotope that transmuted via beta decay into element 93. This first *transuranic* element, meaning "beyond uranium," was named neptunium after the eighth planet (the one following Uranus, for which uranium was named).

The following year a fellow American, Glenn Seaborg, used a cyclotron, like the one used in the creation of technetium, to bombard uranium with deuterium nuclei. (Deuterium is a heavy isotope of hydrogen with a nucleus of one proton and one neutron.) The result was element 94, named plutonium for Pluto, then considered to be the ninth planet. Seaborg went on to develop the means to manufacture plutonium on a large scale for the Manhattan Project. The first atomic explosion (the Trinity test in New Mexico) and the Fat Boy bomb that hit Nagasaki, Japan, in 1945 came from plutonium produced mainly at the Hanford Site in Washington.

Seeing stars

The most stable isotope of neptunium has a half-life of two million years, while plutonium's is ten times as long. Therefore it is possible that these two elements had been present in Earth's rocks in the distant past but have now decayed out of existence. Astronomers applied atomic physics to stars to explain how nuclear fusion turns hydrogen into helium and then other common elements up to iron. But then the physics draws the process to a halt. Heavier elements, astronomers say, must be made by the immense forces of a supernova, in which a giant star explodes so violently that light atoms smash together to make more massive atomic nuclei.

Smashing atoms

In 1944 Seaborg and others, such as Albert Ghiorso, used a cyclotron to smash all kinds of nuclei and other atomic particles together to create ever more exotic substances. The cyclotron used magnetic and electrical fields to direct charged particles in a spiral, so they hit a central target at great speed—enough to cause a nuclear reaction. Over the decades Seaborg created more than 100 new isotopes, totaling another nine new elements. (Albert Ghiorso, Seaborg's co-discoverer on several occasions, continued the work and went on to discover a total of 12 new elements, a world record.) By 2012, 26 transuranic elements had been identified.

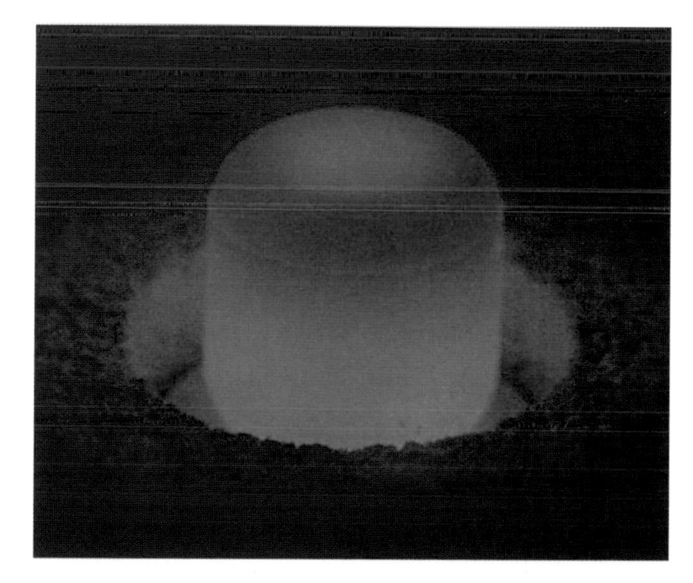

A pellet of plutonium 238, illuminated by the glow of its own radioactivity.

92 The Miller-Urey Experiment

The neighbors coming for a visit?

PANSPERMIA

Several great scientists, Svante Arrhenius and Fred Hoyle among them, have suggested that life arrived on our planet inside a rocky asteroid or comet. Variations of the theory suggest that life was seeded on Earth by anything from fully formed cells to complex biochemicals—that continue to spread across the Universe.

AFTER THE MASS DESTRUCTION OF THE SECOND WORLD WAR, SO MUCH OF IT FORGED BY DESPERATE research into physics and chemistry, many of the scientists involved took the opportunity of peacetime to investigate life itself.

Harold Urey was one of the prime movers in the Manhattan Project, developing gas diffusion techniques for enriching nuclear explosives. He was already a Nobel Laureate after discovering heavy water in 1934. (Heavy water is D_2O, made with deuterium, an isotope of hydrogen with an atomic mass of two instead of one.) After the war he turned his proven intellect to the heavens and began to study the chemical make-up of planetary atmospheres, and to compare and contrast how they might have formed and changed over time. He suggested that Earth's early atmosphere was devoid of free oxygen but rich in water vapor, methane, ammonia, carbon dioxide, and hydrogen. Stanley Miller, one of Urey's graduate students, suggested that these chemicals were the raw ingredients of the so-called building blocks of life—proteins, fats, and carbohydrates,

Cooking the soup

In 1953, the pair decided to test this hypothesis by creating a microcosm of the primordial conditions on Earth. They sealed gases and liquids into a complex network of bulbs and jars that heated, cooled, stirred, and electrified the contents at various points, so the mix was constantly being boiled, condensed, and

CRADLES OF LIFE?

The superheated water of a hydrothermal vent harbors unusual forms of microscopic life. While it is inhospitable to most modern organisms—it would cook them in seconds—this kind of habitat was probably the most stable place on the young Earth, which was frequently rocked by comet strikes and tumultuous volcanic events. Stability is what life needs to evolve, and biochemists think that the warm, chemically rich rock crevices in the seabed are where life is most likely to have started.

given the energy to react with whatever was around it.

They left the apparatus running constantly. Within a day the clear mixture had gone pink, and then within a week more than ten percent of the carbon atoms had formed into organic compounds such as amino acids. Amino acids are monomers, units that chain together to make proteins. There are about 20 amino acids that are used in organisms. Miller announced that the experiment had produced 11 of them. In fact, modern analysis of the apparatus revealed that 20 amino acids had been formed in the seething multitude of reactions that occurred within.

Reruns of the experiment included nitric oxides and sulfur compounds that would have been erupting from volcanoes, resulting in an even larger haul of biologically active chemicals. Although the chemical processes were still untraced, Miller and Urey had shown that life could have been born out of the simplest compounds.

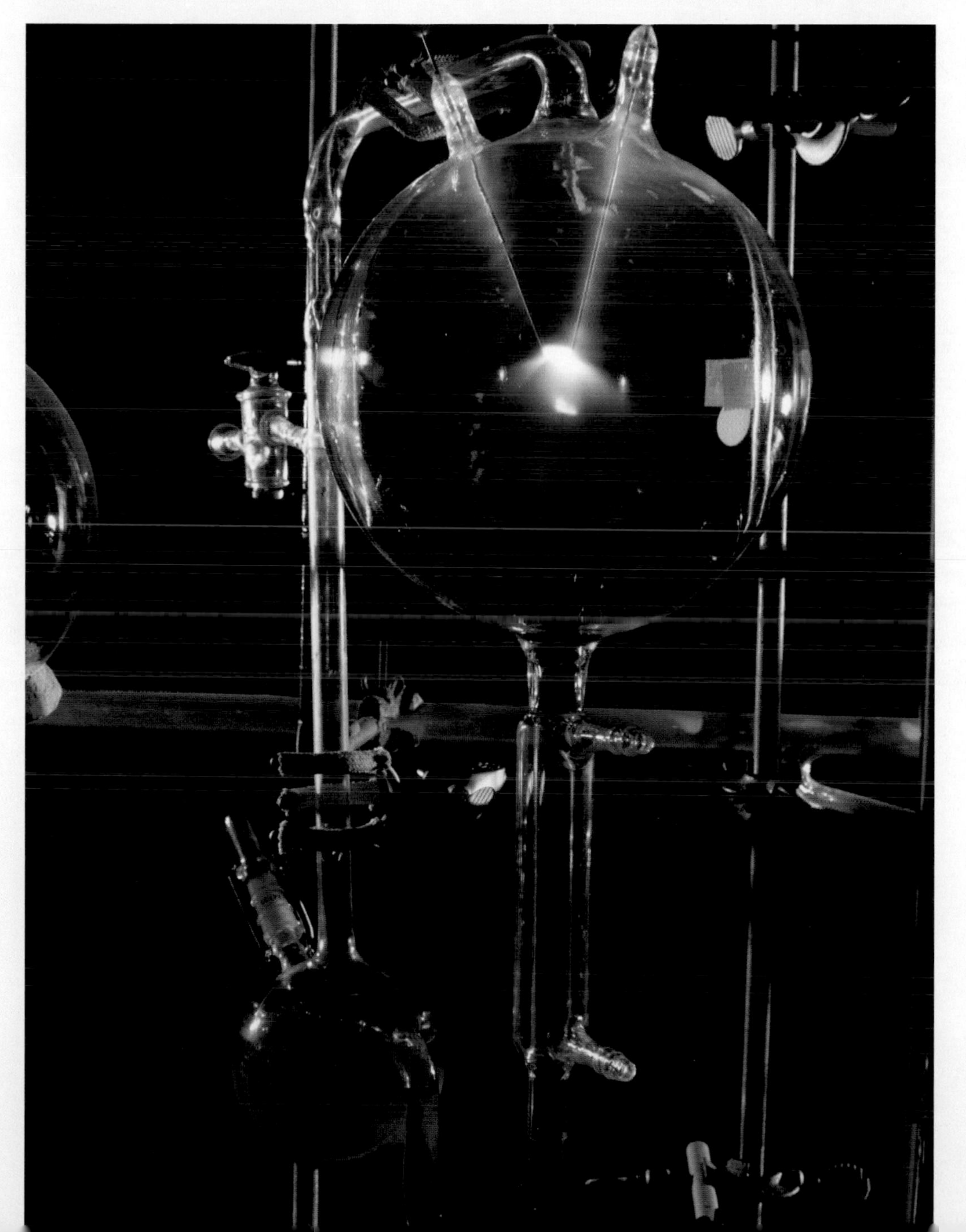

The Miller-Urey experiment has been repeated many times, including with updated apparatus such as this 1983 version.

93 DNA Code

ONE OF THE MOST COMPLEX YET REVEALING PUZZLES TACKLED BY BIOCHEMISTRY WAS FINDING THE STRUCTURE OF DNA. With it came a whole new science.

James Watson and Francis Crick show off a model of the DNA double helix in 1953.

The theory of evolution as set out by Charles Darwin in 1869 made reference to an as-yet unknown genetic material that was passed from generation to generation, carrying with it inherited characteristics of genes. In the same year, a substance found only

in the nuclei of cells was discovered. Shortly this so called nuclein was revealed to contain a ribose sugar component, phosphates, and five aromatic "base" compounds. Together, they formed into long chains of a nucleic acid named deoxyribonucleic acid, or DNA. In 1928 it was becoming clear that DNA was the very genetic material heralded by Darwin. But how did its relatively simple constituents work together to transmit information?

The double helix

Several research teams approached the problem. Even the great Linus Pauling was searching for the structure. The most reliable approach was to use X-ray crystallography to calculate DNA's geometry. However, it was painstaking work to map such a large and complex molecule.

At Cambridge University, Englishman Francis Crick, another physicist who had turned to biology after the war, had teamed with American James Watson to model the structure of DNA from whatever evidence was available. In 1952, they received an X-ray image (designated Photo 51) from Maurice Wilkins at King's College, London, which showed that the DNA was two strands arranged in a spiral—a double helix.

The following year Crick and Watson announced a workable structure for DNA, one which revealed how a chemical could carry information. The sides of the ladder-shaped helix were made from ribose, connected together with phosphates. Pairs of bases made up the "rungs." Only four of the five bases identified were used and they always bonded with a specific partner: thymine linked to adenine, while cytosine partnered with guanine. Uracil, the fifth base, replaces thymine in RNA (ribonucleic acid, a related chemical). Crick and Watson's molecule spelled out a code using the four bases as characters, generally represented with the letters CGTA. A gene was a strand of DNA with a unique sequence of bases. The question of how the DNA code translates into a trait like eye color or an inherited disease spawned the modern science of genetics and has still to be fully answered.

MISSING LINK

Photo 51 was taken by Rosalind Franklin, a researcher at King's College, London. It was supplied to Crick and Watson by her supervisor, Maurice Wilkins, without her knowledge. Franklin's work was published in the same journal as the double helix discovery, but she was given little credit by the discoverers. Crick, Watson, and Wilkins received Nobel prizes in 1962, but Franklin had died of cancer four years before.

94 Understanding Enzymes

AN ENZYME IS A BIOLOGICAL CATALYST, cellular machinery that is purpose built for a specific function. There are 4,000 currently known, each one formed from a protein with a structure coded by a gene of DNA.

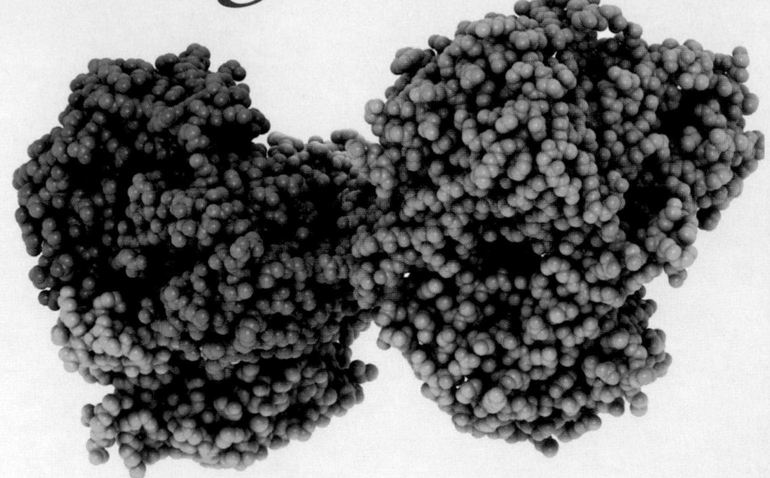

This protein molecule converts seratonin into melatonin, the hormone that controls sleep and wakefulness.

Life is maintained by the chemical action of non-living entities called enzymes. The term was first coined in relation to the action of yeast in brewing and breadmaking and is derived from the Greek for "leaven." However it has since become redefined somewhat to refer to any biological entity that breaks down substances (in catabolism) or builds them up (anabolism).

In 1897, Eduard Buchner showed that enzymes worked just as well outside of a cell or body. Soon it was found that they were all proteins with specific structures and shapes that were central to its metabolic function. If this shape was denatured by heat or the action of chemicals, then the enzyme no longer worked.

LOCK AND KEY HYPOTHESIS

Enzymes cannot multi-task. They can only process one type of material, or substrate. Since as far back as 1894 it has been understood that the shape of the enzyme is specific to its substrate, like a key fitting in a lock. The "lock" section is also called the active site of the enzyme. The substrate or substrates bond to it, and in so doing are chemically changed into products. Many toxins work by bonding to active sites and blocking the action of crucial enzymes.

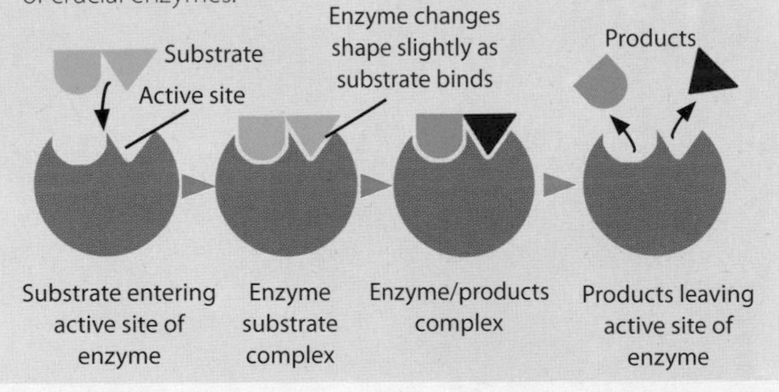

Amino acid chains

A protein is a polymer made up of amino acids, and a typical enzyme contains hundreds, if not thousands, of these individual units. There are about 20 amino acids in use in nature, and the genetic code of DNA is a list of amino acids for enzymes and other proteins. These then do all the other jobs the body needs. The order in which they are arranged in a protein is called its primary structure. In 1965, X-ray crystallographers first identified the structure of an enzyme (lysozyme, used in a digestion), showing that a protein's primary structure results in a secondary one, which is caused by various portions of the long acid chain folding and twisting in on itself. This can lead to tertiary and even quaternary structures. A small change, or mutation to the code, can alter a single amino acid, which can have a significant impact on the structure and function of the enzyme.

95 Buckminsterfullerene

UNTIL THE 1980S, THERE WERE THREE ALLOTROPES, OR FORMS, OF PURE CARBON: DIAMOND, GRAPHITE, AND SOOT. Then an exquisitely complex new form was discovered, where the carbon atoms form perfect spheres of varying sizes. These were the fullerenes, alternatively styled "buckyballs," and they may change technology forever.

Carbon atoms can form up to four bonds with other atoms. Diamond comprises repeating units of five carbon atoms in the shape of a tetrahedron, or triangular-based pyramid. The four corners of this pyramid are formed by four carbon atoms bonded to the fifth at the center. Arranged together these tetrahedra form an immensely strong lattice, which is why diamond is the hardest material on the planet. By contrast graphite is very soft and slippery. Its carbon atoms are arranged in layers of hexagons. Each atom is bonded strongly to three others, while a weaker fourth bond links the atoms in one layer to another above or below, meaning that the layers move over one another easily. This is why graphite is used as a lubricant and in pencil leads, where a layer of carbon is left behind on the page as the pencil tip rubs against the page.

Spherical carbon

A fullerene is equivalent to the hexagonal graphite layer being curved into a ball. Anything between 20 and 100 carbon atoms can achieve this but the most common number is 60. C_{60} was the first fullerene to be identified and has the full name buckminsterfullerene, from which fullerene and buckyball are both derived as general terms for similar molecules.

C_{60} was predicted to exist in 1970 and evidence of it was found in the spectral analysis of dust clouds in deep space. However, it was not synthesized on Earth until 1985, when graphite was vaporized in a helium atmosphere. The helium prevented the carbon from burning. Instead it folded into C_{60} and other spherical molecules. U.S. chemists Richard E. Smalley and Robert F. Curl Jr., together with English chemist Harold Kroto were awarded the Nobel Prize for this work in 1996.

At first fullerenes were something of a curiosity that existed in only tiny amounts in nature, largely in soot or formed by lightning. However, later developments would put carbon and its fullerenes at the heart of a potentially revolutionary technology.

A bucky ball is composed of 20 hexagons and 12 pentagons.

WHAT'S IN A NAME?

Buckminsterfullerene is named after Richard Buckminster Fuller, an American architect who is known best for his geodesic domes, which are lightweight yet sturdy enough to span huge areas. Buckminster Fuller designed the domes in the 1950s, and only later did chemists find that carbon atoms formed precisely the same shapes.

The Montreal Biosphere was designed by Buckminster Fuller for the Expo 67 event.

96 Seeing Atoms

BY THE MID 1980S, THE SCANNING TUNNELING MICROSCOPE WAS DEVELOPED TO IMAGE ATOMS. This new device gave chemists a closer look at their subjects than had ever been thought possible.

The scanning tunneling microscope (STM) can see things that are 0.1 nanometer (nm) across—that is the size of the helium atom. Hydrogen is the only element still out of view. The microscope uses an electrified tungsten probe tapered to a tip a single atom wide. When the tip is brought close enough to a sample, a phenomenon called quantum tunneling occurs, in which a vortex of electrons travels across the divide between the probe and the nearest atom below. These fluctuating connections are used to map the exact position of atoms.

This STM image of a piece of platinum resembles a contour map with every 10-nm atom bulging from the surface into a red peak.

97 High-Temperature Superconductors

A SUPERCONDUCTOR IS A SUBSTANCE THAT CARRIES A CURRENT OF ELECTRICITY WITHOUT OFFERING ANY RESISTANCE. The first superconductors had to be chilled to incredibly low temperatures. But all that changed in 1986.

In 1911, Dutch physicist Heike Kamerlingh Onnes chilled mercury to –269°C, and found that the now frozen metal behaved as a superconductor. The potential was immediately obvious, but the temperature required was beyond the limits of the everyday—and even colder than deep space.

In 1986, Karl Müller and Johannes Bednorz developed a ceramic that could superconduct at only –163°C. This may not seem very high but in the world of superconductors anything more than the boiling point of liquid nitrogen (–196°C) is "high" and therefore relatively easy to maintain. High-temperature superconductors are now commonly used in the electromagnets in MRI scanners, mass spectrometers, and particle accelerators, where they are efficient enough to produce huge magnetic fields.

A maglev (magnetic levitation) train relies on superconducting electromagnets to lift it off the track, so it can glide along at record-breaking speeds free from the hindrance of friction.

98 Nanotubes

SINCE THE 1990S BUCKYBALLS HAVE BEEN REFASHIONED INTO TINY TUBES. SCIENTISTS ARE LEARNING TO GROW tubes to any length and a range of widths. This raises many possibilities, from nano-sized machines to high-speed superconducting communication cables. The future looks tube-shaped.

Sumio Iijima's invention is the narrowest tube ever constructed. A nanotube long enough to stretch to the Moon would roll up into a ball no larger than a poppy seed.

Carbon nanotubes were first prepared by Japanese industrial chemist Sumio Iijima in 1991. A nanotube can be imagined as a buckyball in which extra rings of hexagons have been added, creating a sausage shape. The current manufacturing process is to roll them from graphene, a new term used to describe a layer of graphite (also comprised of hexagons) that is just one atom thick. Metal catalysts are used to stop the tube end, closing it off with pentagonal faces, but the hope is that tubes of unlimited length will be possible. By contrast the widths of nanotubes are measured in angstroms. A single angstrom is a ten billionth of a meter.

Nanotechnology

As in benzene, the fourth bond of all the carbon atoms is shared, forming a cloud of charge that gives the fullerene and nanotube its stability. However, unlike benzene, nanotubes will conduct electricity, more than 1,000 times as much as copper wires. This means that one day nanotubes will replace all metal wires and even optical fiber telecommunication cables. In addition, the tubes can be made to act as semiconductors, promising a whole new design of even smaller integrated circuits and computer technology.

For their size nanotubes are very strong, thousands of times stronger than steel. Perhaps bundles of nanotubes will be used to build longer bridges or super-lightweight vehicles. They could also be constructed into machinery thousands of times smaller than is currently possible.

Tubes within tubes could be used as superconductors, offering zero resistance to current.

99 An Island of Stability

THE HEAVIER TRANSURANIC ELEMENTS BECOME, THE LESS STABLE THEY ARE, AND OF LESS USE. However, it is predicted that there is an "island" of stable atoms in the periodic table.

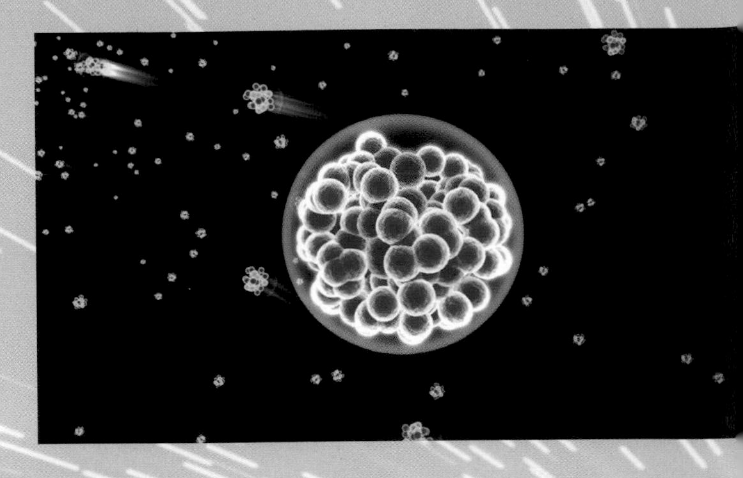

The largest atom that has currently been formed is number 118 (temporarily named ununoctium), It has a half-life of 0.89 milliseconds, so it does not hang around very long. However, Glenn Seaborg suggested that atoms with larger numbers of particles in the nucleus would have more complete shells and be more stable. Their half-lives would be at least days, if not millions of years. The most stable is predicted to be 126 which would have a uniquely complete set of protons and neutrons. The island of stability would come to an end at element 137.

In 2010, ununseptium (117) became the latest element to be discovered. The island of stability begins at element 120.

100 The Higgs Boson

IN 2011, SCIENTISTS BECAME HOPEFUL THAT THEY WOULD GET AN ANSWER TO ONE OF THE BIG QUESTIONS IN SCIENCE: DOES THE HIGGS BOSON EXIST? To the layman this question may lack poetry but it has certainly caught the public imagination, not least because the world's scientists have built the largest machine in history to answer it.

When J.J. Thomson discovered the electron in 1897, he was making the first entry into what is now known as the Standard Model. This is a family of particles that can be used to explain matter—at least the matter that we can currently observe—and unifies three of the four fundamental forces that act on matter. Those forces are the strong interaction, which holds an atomic nucleus together; the weak interaction, which pushes particles out of the nucleus during radioactive decay; and electromagnetism, which holds electrons around the nucleus and is also involved in chemical bonding, electricity, and the push and pull of magnets. The force of gravity is the only one unaccounted for so far—assuming we do not discover more fundamental forces.

A force is a transfer of energy from one mass to another, and this is achieved by a particle carrying it between the masses. Particles that mediate force are called bosons, so the photon is the boson for electromagnetism. The weak force is mediated by the W and Z bosons, while the strong force at the heart of the atom is carried by gluons.

Where did mass come from?

Bosons move between masses, which are any objects that are affected by forces. Objects can be as large as a galaxy or as small as an electron. A galaxy is made up of stars and planets constructed of atoms, including their electrons. In the 1970s it was found that protons and neutrons were in fact made up of trios of smaller particles called quarks—this is what the gluons are holding together. But one of the Standard Model's unanswered questions is what makes electrons and quarks (and the other particles) have mass. The suggested answer is that the Universe is filled with a field, the Higgs field (named for English physicist Peter Higgs). It is this field's boson particle—the Higgs boson—that gives mass to particles in it.

The theory suggests that the Higgs field was not formed at the same time as the Big Bang, but a short time after. This theory will be tested by the world's scientists at the Large Hadron Collider (LHC), at CERN in Switzerland. This is the most powerful particle accelerator ever built, and is smashing protons together at nearly the speed of light to recreate the intense energy that existed just after the Big Bang. Scientists then wait for evidence of a Higgs field forming. If they find any, it will not be the final question answered, just the first glimpse of new information on how the Universe works.

A BIG BANG

Since the Universe is expanding now, it must have been smaller in the past. Just shy of 14 billion years ago it took up no space at all, before creating space itself and going off in a Big Bang. This was an event that took place everywhere at the same time, it was just that everywhere would have fitted inside a grapefruit. The Big Bang theory was named as a disparaging joke by English astronomer Fred Hoyle who was not a fan of the concept. However, his term stuck.

Energy became mass, but did it involve the Higgs boson?

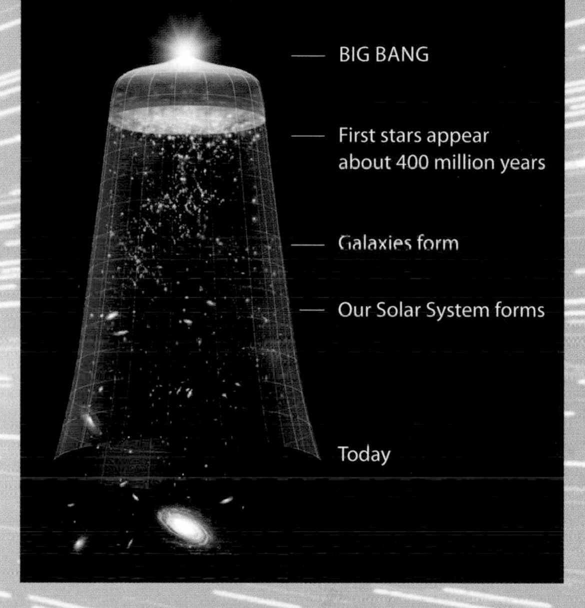

The Large Hadron Collider is in a 17-mile (27-km) tunnel that connects various detectors. Getting around to service the 8,000 superconducting magnets that control its beam of particles is not always easy.

101 Chemistry: **the basics**

SO WHAT DOES ALL THIS DISCOVERY ADD UP TO? If we take a look at the elements from another angle, drawing all the lines of inquiry together, we find that the elements are the very basis of chemistry itself.

What is an element?

The Universe is made from a set of building blocks that combine to produce all the substances around us, from our planet's water, rocks, and air to the burning furnace of a star. These building blocks are the chemical elements; there are 92 found on Earth, although most of them are extremely rare. Common elements include oxygen, carbon, silicon, and iron. What sets elements apart from other materials is that they cannot be broken down into any simpler substances.

Few elements are found pure in nature: gold (below left) is one of these so-called native substances. Most elements are chemically combined with each other and have to be be refined into a pure form, such as this iron (below).

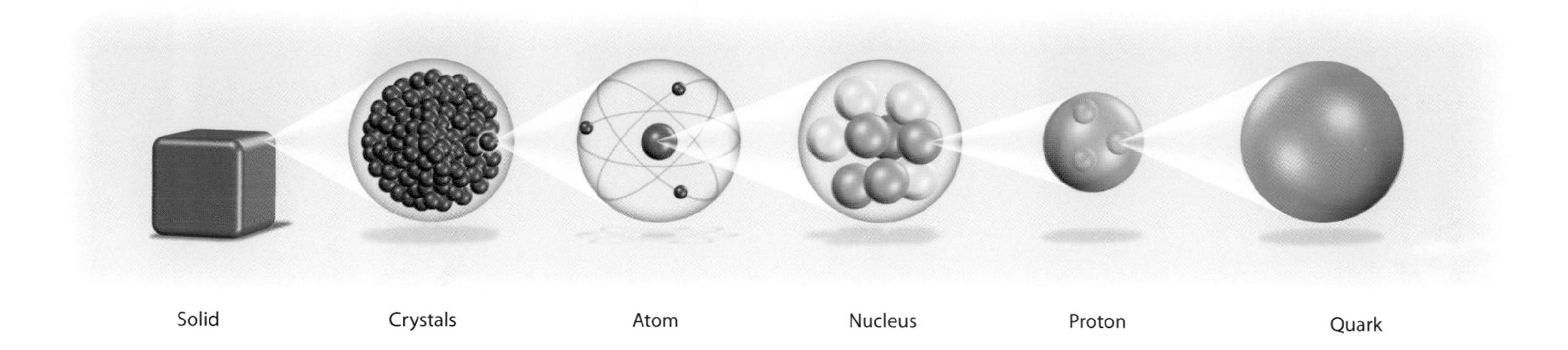

Solid Crystals Atom Nucleus Proton Quark

Matter can be described at a number of scales, or frames of reference. On the macroscale it has color and form (a solid in this case). Next comes the molecular structure, the way the atoms combine to form crystals or other forms. Then there is the atom itself, with most of its matter in the nucleus. The particles in the nucleus—the nucleons (protons or neutrons)—are themselves made up of three quarks.

What is an atom?

The smallest unit of an element is an atom. Give or take a few, 33 million would line up to just one millimeter in length. A pure sample of an element comprises just one type of atom, with a specific size, mass, and structure. It is these properties that give every element its unique suite of characteristics. This is why oxygen is a gas while gold is a glinting, yellow metal.

While an element cannot be simplified beyond the level of an atom, atoms are themselves composed of smaller particles, such as electrons, protons, and neutrons. The form of these subatomic particles is not specific to the atoms of different elements, so as an example, the electrons in hydrogen are interchangeable with those in chlorine. It is the quantity of particles in an atom that define which element it belongs to.

Atomic structure

Every element has a specific atomic number. This is the number of protons it contains, which are located in the nucleus. Hydrogen has one proton, helium two, and so on all the way to uranium, which has an atomic number of 92, the highest of any naturally occuring element. The protons carry a positive electric charge, but atoms are neutral overall because that charge is balanced by an equal number of negatively charged electrons. The protons reside in the atomic nucleus, while the electrons are arranged in layers, or shells, around it. Each electron shell has room for a specific number of electrons. When full, a new shell begins further from the nucleus. As a result some atoms have outer shells that are nearly full of electrons or nearly empty, and this configuration is the dictating factor in how the atom forms bonds.

The electron configuration of the first ten elements showing how electrons are added to atoms as the atomic number increases. This configuration is reflected in the arrangement of the periodic table. With the row, or period, corresponding to the number of electron shells an atom has.

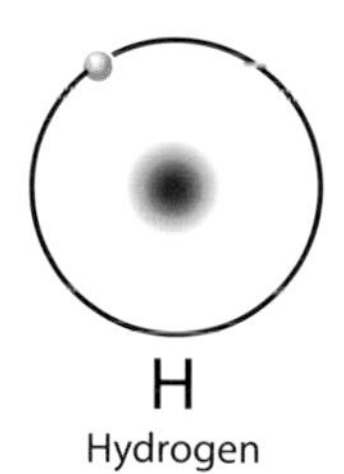

H
Hydrogen

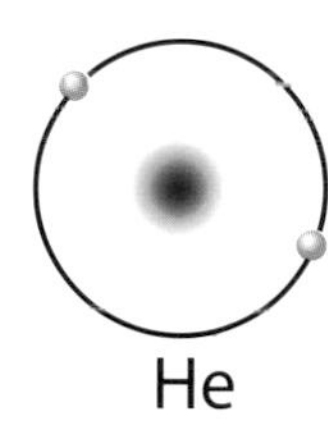

He
Helium

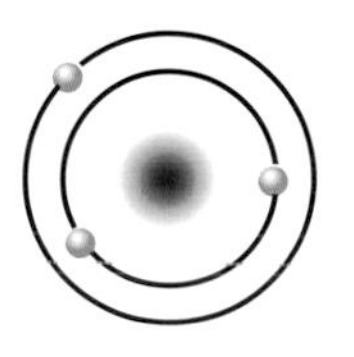

Li
Lithium

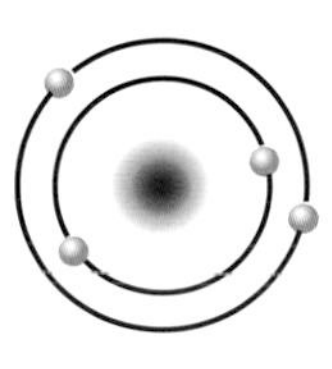

Be
Berylium

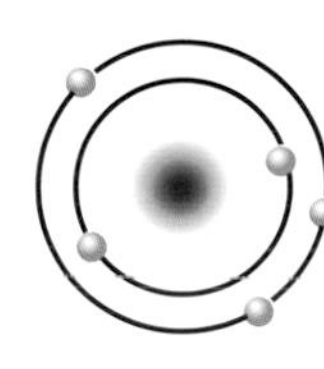

B
Boron

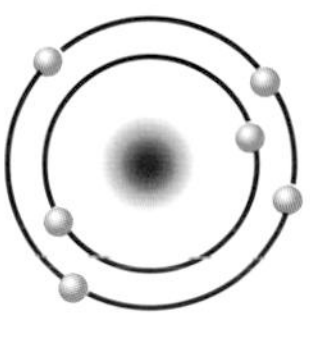

C
Carbon

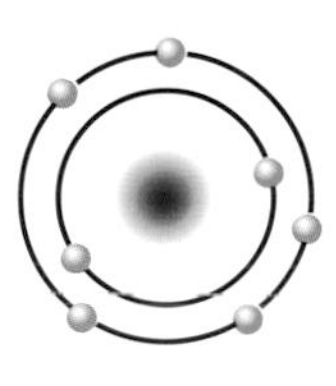

N
Nitrogen

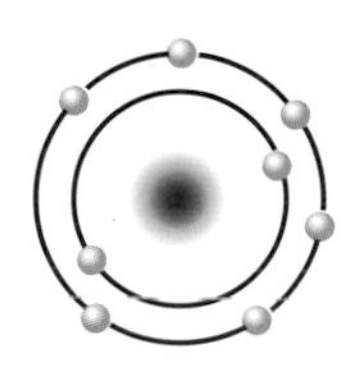

O
Oxygen

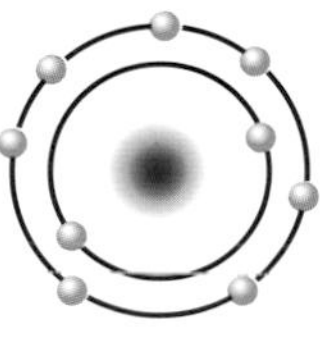

F
Fluorine

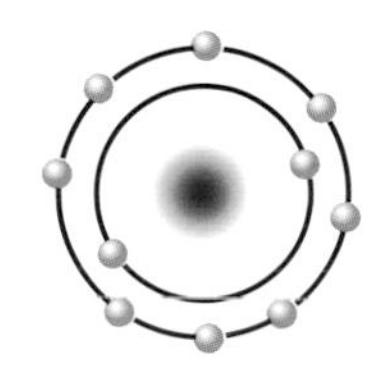

Ne
Neon

Chemical bonds

Ionic bonding

This kind of bond is an attractive force between two ions. An ion forms when an atom loses or gains electrons, unbalancing its charge. Atoms with just a few outer electrons (such as those of metals) tend to lose electrons and form positive ions, while non-metal atoms tend to gain electrons to fill the few remaining holes in their outer shell. Ions with opposite charges attract, while like charges repel.

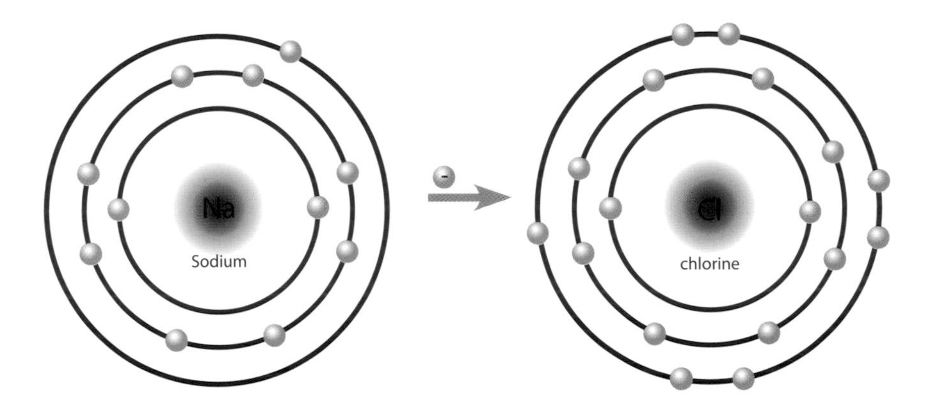

A sodium atom loses its single outer electron to become a positively charged ion; chlorine takes the electron to fill its outer shell and so forms a negative chloride ion. The two ions bond forming a molecule of sodium chloride, or common salt.

Covalent bonds

Atoms ionize and form ionic bonds in order to achieve a stable, low-energy state. As a result those elements that have electrons to lose and those that have them to gain will bond in his way. However, atoms can also form stabilizing bonds not by exchanging electrons, but by sharing them. This is a covalent bond, where two or more atoms come together so their outer electron shells—the valence shell—cross over. As a result an electron from one atom pairs up with one from another, and the two occupy the valence shells of both atoms at the same time. The negatively charged electrons are being held in place by the positive charge of a nucleus, and the pull of two nuclei on the paired electrons keeps the atoms together. Covalently bonded structures can go on to lose or gain electrons, forming large polyatomic ions with one overall charge.

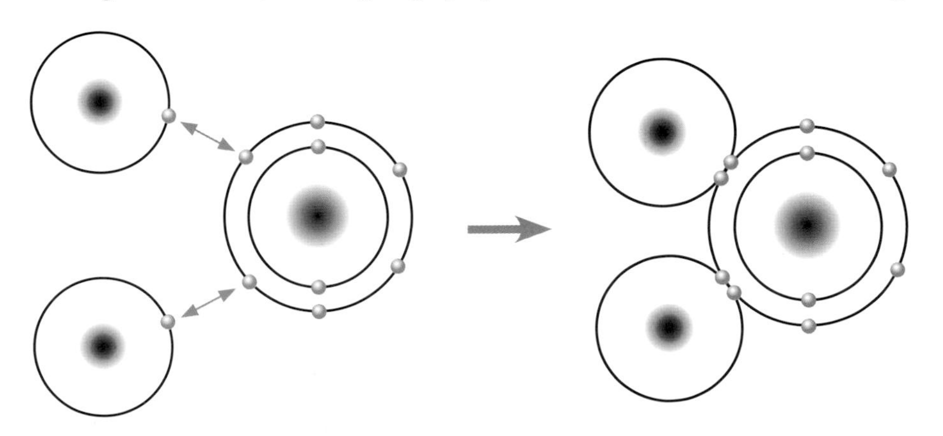

Covalent bonds are frequently made between two non-metal atoms. Here two hydrogen atoms are bonding to an oxygen to form a molecule of water.

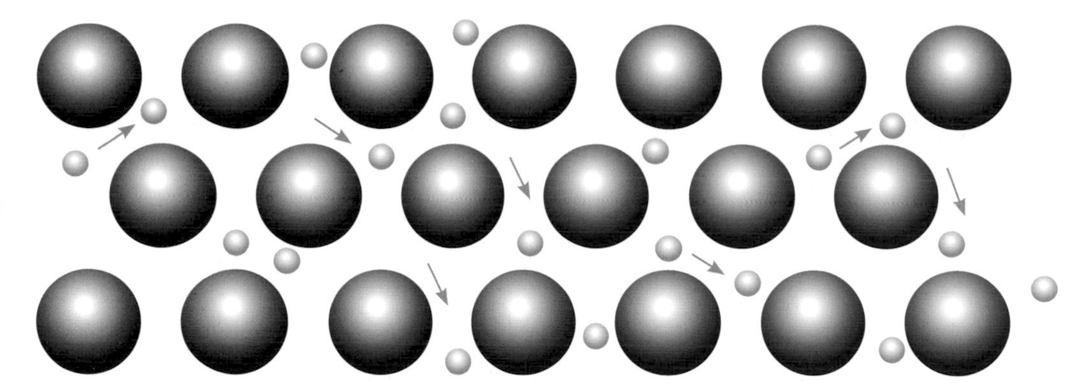

A metal object—either pure or several metals alloyed together—is filled with free electrons. If these electrons can be made to travel in the same direction by an electromagnetic force it results in an electrical current.

Metallic bonds

What makes an element a metal? Metals tend to be shiny, dense solids that can be hammered or molded into shapes, and this is due to their electron configurations. Most metals have just one or two outer electrons—only a handful have three or four. These outer electrons are lost easily, which is why metals form ionic bonds with non-metals, elements that collect electrons rather than lose them. Metallic atoms also give up their outer electrons when bonding to each other. This creates a "sea" of shared electrons that surround the atoms, forming a subatomic glue that binds them together. These strong metallic bonds keep metals intact even when buckled and bent.

Reactions, compounds, and molecules

As a quick look around will confirm, pure elements are rather rare. It would appear that when left to their own devices atoms prefer to band together in clusters. Even the elemental gases in air—mainly oxygen and nitrogen—do not exist as single atoms. Instead, two oxygen atoms bond together covalently to form a structure called an oxygen molecule (O_2).

Just as atoms are the smallest units of an element, so molecules are the smallest unit of a compound. A compound is a substance that is made up of the atoms of two or more elements bonded together in specific proportions to produce a molecule with a particular shape. A compound need not resemble its parent elements—and seldom does. Water is a compound of two gases, hydrogen and oxygen. The salt we season food with is a compound of an explosive metal and acrid green gas.

Compounds are formed by chemical reactions. In simple terms there are three types of reaction: synthesis, decomposition and displacement. A

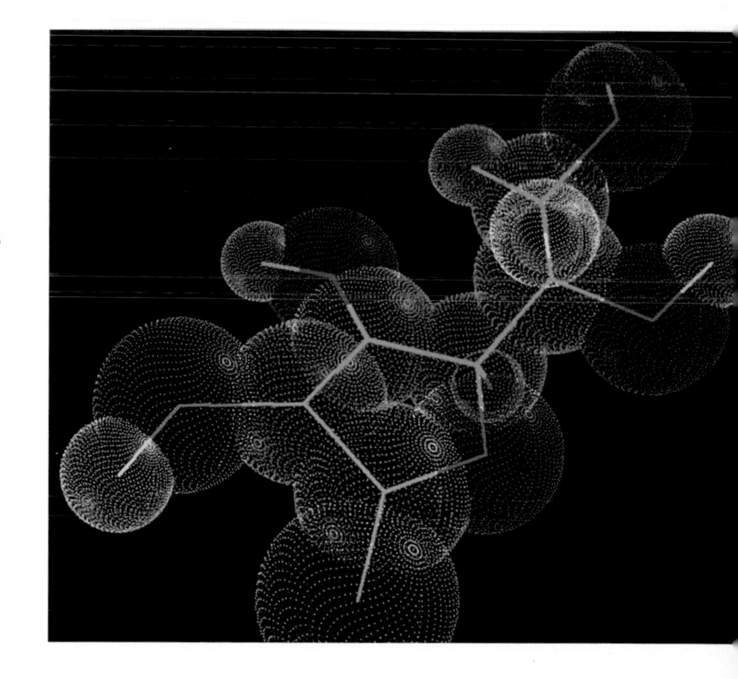

Vitamin C, or ascorbic acid, is a compound of carbon (green), oxygen (red), and hydrogen (white) atoms arranged in a complex molecular structure.

synthesis is when two reactants combine to form a single product, such as carbon and oxygen forming carbon dioxide. A decomposition is when a single reactant splits into two products, such as when carbonic acid dissociates into water and carbon dioxide (adding the sparkle to a soda). Finally a displacement is when a more reactive element substitutes a less reactive one.

Groups and trends

The chemical properties of elements, especially the proportions in which they form compounds with other elements, makes it possible to group them together. This is one of the ways the elements are organized on the periodic table. It is now known that the chemical characteristics of these groups is due to their similar electron configurations. For example, Group 1 elements have one outer electron, while those in Group 2 have two etc.

Members of a group are not all identical. There are always exceptions, such as the liquid metal mercury, one of only two elements that are liquid in standard conditions,

Mercury is a dry liquid. Its atoms bunch together so much that they do not spread over surfaces to make the surface wet. Instead it stays in blobs.

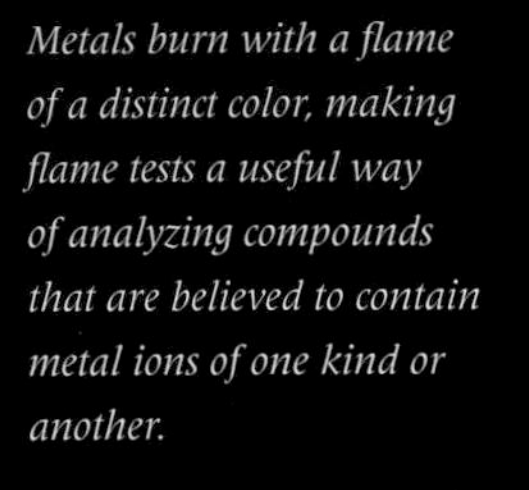

Metals burn with a flame of a distinct color, making flame tests a useful way of analyzing compounds that are believed to contain metal ions of one kind or another.

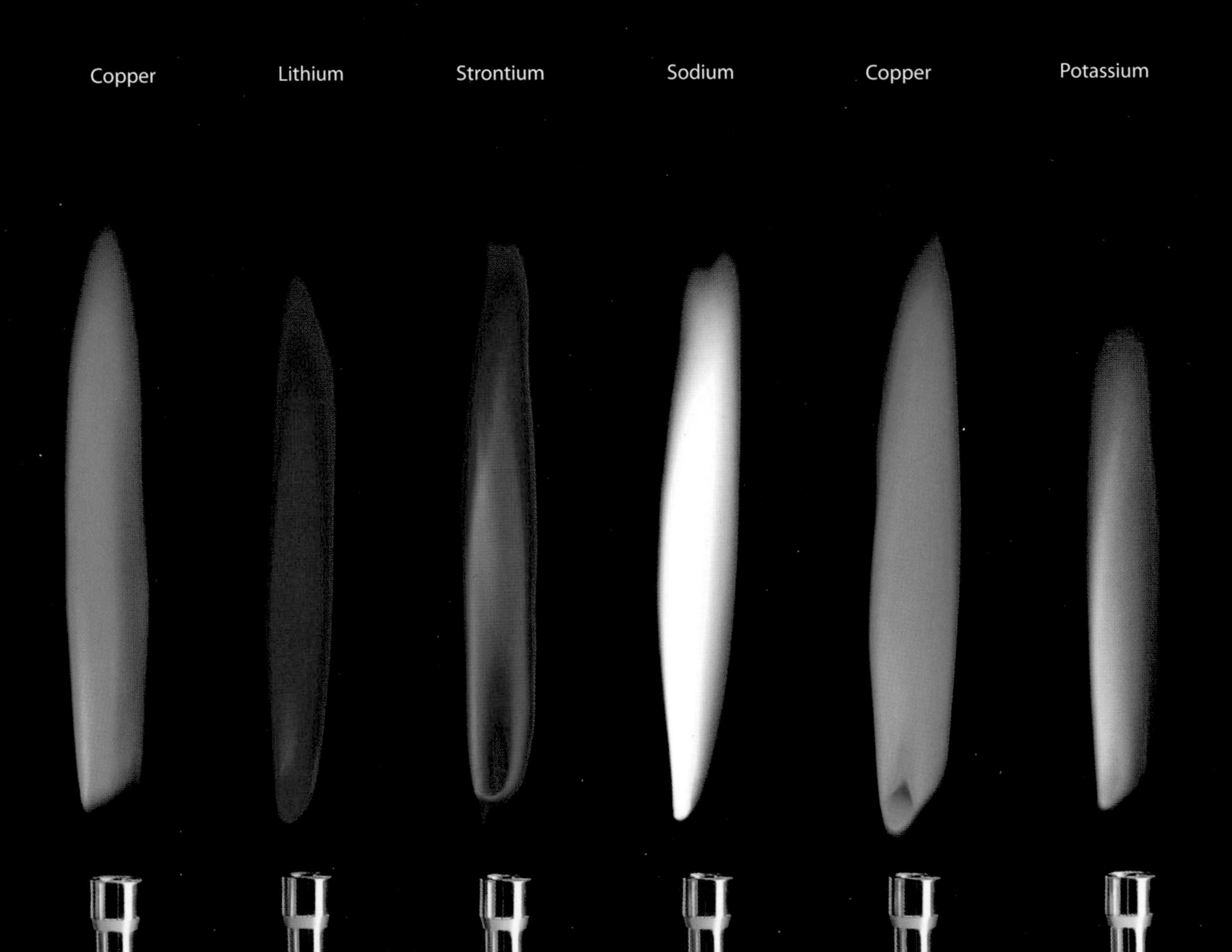

but it is possible to get a good idea on how an element is likely to react by checking which group it is in on the periodic table. Metals are on the left, while non-metals are on the right. Elements from either side have opposite requirements in order to achieve stability and so they react with each other easily.

Metal atoms give away electrons easily, and those with large atoms that have their outer shells located far from the pull of the nucleus can hold their outer electrons only loosely. As a result they form ions and become involved in reactions very easily. In other words, heavy members of a metallic group are more reactive than the lighter ones. The opposite is true of non-metals which react by taking in extra electrons. Smaller atoms pull on these additional electrons more strongly than their larger, heavier cousins and are more reactive as a result. Fluorine and cesium make a big bang!

Ice

States of matter

All matter, both pure elements and compounds, has three basic forms: solid, liquid, and gas. A solid like ice has a fixed shape; a liquid, such as water, takes on the shape of its container and can flow from one vessel to another; a gas, like water vapor or steam (hot vapor), will also take on the shape of its container but will spread out to fill the entire volume.

Water

Every substance has a standard state but it can be changed from one state to another by adding or taking away heat or changing the pressure it is under (squeezed gas condenses). The atoms and molecules inside a substance are always moving, even those held within a solid. Heat is a measure of how much movement there is. When the solid is heated, the molecules begin to vibrate more strongly until the force of their motion is greater than the strength of the bonds holding them together the bonds begin to break. Once about 10 percent of a solid's bonds break, it melts into a liquid. The molecules are still largely bound together but are able to move past each other, and that allows the liquid to flow. If the liquid is heated further, the molecules begin to vibrate even more quickly. Eventually they are vibrating so much that all the bonds between the molecules break, and the liquid turns into a gas. As a gas, the molecules are free to move in any direction. The molecules hit other gas molecules or bounce off the solid surface of a container, but soon spread out evenly. Removing energy from the molecules and reducing their motion will run the whole process in reverse.

Steam

The Periodic Table

Alkali metals

Alkaline-earth metals

Transition metals

Metaloids

Poor metals

Other non-metals

Halogens

Noble gases

Lanthanides

Actinides

Radioactive

1	2	3	4	5	6	7	8	9
1 H HYDROGEN 1.00794								
3 Li LITHIUM 6.941	4 Be BERYLLIUM 9.012182							
11 Na SODIUM 22.98976928	12 Mg MAGNESIUM 24.305							
K POTASSIUM 39.0983	Ca CALCIUM 40.078	21 Sc SCANDIUM 44.955912	22 Ti TITANIUM 47.867	23 V VANADIUM 50.9415	24 Cr CHROMIUM 51.9961	25 Mn MANGANESE 54.938045	26 Fe IRON 55.845	27 Co COBALT 58.933195
Rb RUBIDIUM 85.4678	Sr STRONTIUM 87.62	39 Y YTTRIUM 88.90585	40 Zr ZIRCONIUM 91.224	41 Nb NIOBIUM 92.90638	42 Mo MOLYBDENIUM 95.96	43 Tc TECHNETIUM 98	44 Ru RUTHENIUM 101.07	45 Rh RHODIUM 102.9055
Cs CAESIUM 132.9054519	Ba BARIUM 137.327	57-71 LANTHANOIDS	72 Hf HAFNIUM 178.49	73 Ta TANTALUM 180.94788	74 W TUNGSTEN 183.84	75 Re RHENIUM 186.207	76 Os OSMIUM 190.23	77 Ir IRIDIUM 192.217
Fr FRANCIUM (223)	Ra RADIUM (226)	ACTINOIDS	104 Rf RUTHERFORDIUM 267	105 Db DUBNIUM 270	106 Sg SEABORGIUM 271	107 Bh BOHRIUM 274	108 Hs HASSIUM 277	109 Mt MEITNERIUM 278

57 La LANTHANUM 138.90547	58 Ce CERIUM 140.116	59 Pr PRASEODYMIUM 140.90765	60 Nd NEODYMIUM 144.242	61 Pm PROMETHIUM 145	62 Sm SAMARIUM 150.36	63 Eu EUROPIUM 151.964
89 Ac ACTINIUM (227)	90 Th THORIUM 232.03806	91 Pa PROTACTINIUM 231.03588	92 U URANIUM 238.02891	93 Np NEPTUNIUM 237	94 Pu PLUTONIUM 244	95 Am AMERICIUM 243

Au *Solids*

He *Gases*

Br *Liquids*

Mt *Synthetics*

10	11	12	13	14	15	16	17	18
								2 He HELIUM 4.002602
			5 B BORON 10.811	6 C CARBON 12.0107	7 N NITROGEN 14.0067	8 O OXYGEN 15.9994	9 F FLUORINE 18.9984032	10 Ne NEON 20.1797
			13 Al ALUMINUM 26.9815386	14 Si SILICON 28.0855	15 P PHOSPHORUS 30.973762	16 S SULFUR 32.065	17 Cl CHLORINE 35.453	18 Ar ARGON 39.948
28 Ni NICKEL 58.6934	29 Cu COPPER 63.546	30 Zn ZINC 65.38	31 Ga GALLIUM 69.723	32 Ge GERMANIUM 72.64	33 As ARSENIC 74.9216	34 Se SELENIUM 78.96	35 Br BROMINE 79.904	36 Kr KRYPTON 83.798
46 Pd PALLADIUM 106.42	47 Ag SILVER 107.8682	48 Cd CADMIUM 112.411	49 In INDIUM 114.818	50 Sn TIN 118.71	51 Sb ANTIMONY 121.76	52 Te TELLURIUM 127.6	53 I IODINE 126.90447	54 Xe XENON 131.293
78 Pt PLATINUM 195.084	79 Au GOLD 196.966569	80 Hg MERCURY 200.59	81 Tl THALLIUM 204.3833	82 Pb LEAD 207.2	83 Bi BISMUTH 208.9804	84 Po POLONIUM 209	85 At ASTATINE 210	86 Rn RADON 222
110 Ds DARMSTADTIUM 281	111 Rg ROENTGENIUM 281	112 Cn COPERNICIUM 285	113 Uut UNUNTRIUM 286	114 Uuq UNUNQUADIUM 289	115 Uup UNUNPENTIUM 289	116 Uuh UNUNHEXIUM 291	117 Uus UNUNSEPTIUM 294	118 Uuo UNUNOCTIUM 294

64 Gd GADOLINIUM 157.25	65 Tb TERBIUM 158.92535	66 Dy DYSPROSIUM 162.5	67 Ho HOLMIUM 164.93032	68 Er ERBIUM 167.259	69 Tm THULIUM 168.93421	70 Yb YTTERBIUM 173.054	71 Lu LUTETIUM 174.9668
96 Cm CURIUM 247	97 Bk BERKELIUM 247	98 Cf CALIFORNIUM 251	99 Es EINSTEINIUM 252	100 Fm FERMIUM 257	101 Md MENDELEVIUM 258	102 No NOBELIUM 259	103 Lr LAWRENCIUM 262

IMPONDERABLES

THERE ARE STILL A LOT OF UNSOLVED MYSTERIES IN CHEMISTRY, IMPONDERABLE PROBLEMS STILL TO BE PUT TO RIGHTS. The periodic table is by no means complete, and there are still an untold number of compounds to be synthesized. With the predicted rise of nanotechnology and quantum computing, chemistry still has a future.

Did buckyballs bring life to Earth?

Is it a compound? Is it a mixture? Scientists investigating what you can and can't do with fullerenes have found that they can "inflate" a buckyball with other atoms. In other words the atoms are contained within the ball of 60 carbons. This complex arrangement has been termed a endohedral fullerene. The first was lanthanum C_{60}, which is now represented with the shorthand of La@C60. Since then ions and molecules have been trapped inside the nanoscopic balls. The technological impact of endohedral fullerenes has yet to become clear. However, in 2010 fullerenes were discovered floating around in deep space. That begged the question, could these chemicals hold materials inside as well? And if they do, could this be the mechanism for panspermia—the theory that life was seeded on Earth by chemicals arriving from space? Did life arrive in tiny capsules made from buckyballs?

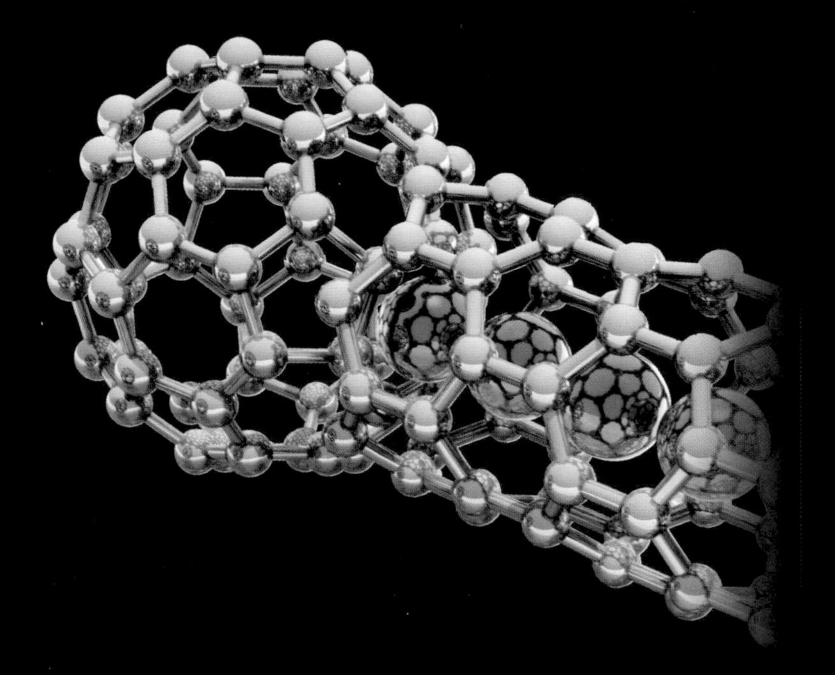

Is bismuth radioactive?

Nuclear physicists have long theorized that bismuth was not the last stable element but was in fact the first radioactive one. In 2003 it was demonstrated that bismuth 209, the only naturally occurring isotope of this heavy metal, did break down by alpha decay, but it did it so slowly that bismuth's half life is a billion times longer than the current age of the Universe.

Does the periodic table run out?

Most of the transuranic elements are extremely unstable, with half-lives of just a few milliseconds. That means that after all the hard work and energy spent on making a few atoms of these new elements, they decay into something else almost immediately. However the "island of stability" predicted between atomic numbers 117 and 127 may change all that, with huge atoms that are predicted to be relatively stable, although still highly radioactive. Then at number 137 the whole system begins to fall apart. At this point the force needed to hold the atom together is beyond what electromagnetism can provide. The hypothetical element 137 is named feynmanium (Fy) in honor of Richard Feynman, the quantum physicist who pointed out this problem. Does this spell the end of the periodic table? Not necessarily the end, but certainly the beginning of the end. An atom of feynmanium could form but would not be neutral, and would thus break the rules governing all other elements to that point. And by element 173, the rule book is in tatters, with atoms predicted to emit a perpetual supply of positrons—positively charged electrons—which are forged spontaneously by the huge forces deep within.

Will silicon nanotubes replace coal?

Like carbon, silicon has a valence of four, meaning it can bond to four other atoms. Silicon–hydrogen chemistry mirrors that of hydrocarbons, although it is considerably less complex. Compounds called silanes, which are analogous to the alkanes such as methane, butane, and octane, are already used in water-repelling paints and windshield coatings. Although silicon buckyballs are too unstable to form, hexagonal cages of silicon, housing a heavy metal atom, have been manufactured. It has been suggested that these units could be used as qubits—quantum bits—where the spin state of the central atom could be used to store data in a high-speed computer of the future. In 2006 nanotubes made from silicon were made for the first time. The possibilities of a semiconducting nanotube have yet to be explored, but it has also been suggested that a matrix of these nanoscale tubes could form a highly porous solid. Such a substance could be used to store hydrogen gas to form a high-density solid fuel, akin to coal. When burned, this material would produce large amounts of heat; sand instead of ash; and zero carbon emissions.

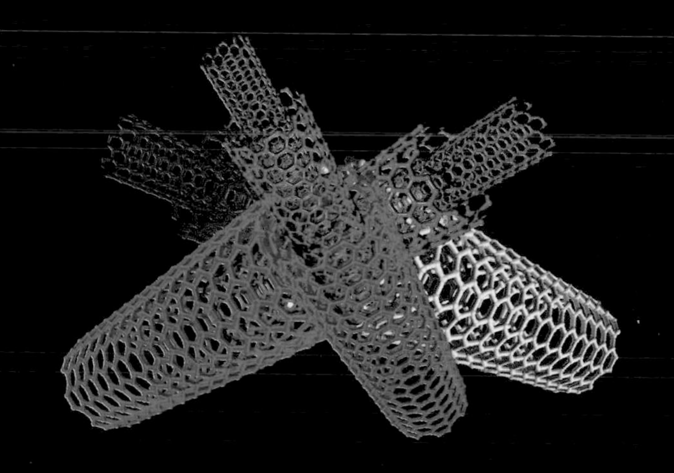

Why is nature one-sided?

As Louis Pasteur discovered, many of the chemicals used to make living things, such as amino acids and sugars, are homochiral: in nature they exist in only one form of the possible mirror images of their isomers. In addition, the metabolism of living things can only handle homochiral ingredients to create proteins, DNA, and other vital substances. Since the opposite-sided version has the exact same chemical constituents and properties, it remains a mystery as to why nature only produces one version of these chemicals. It is theorized that all current biological activity was born out of a more primitive chemical process. The simplest life was based on chemicals that already existed, rather than manufacturing everything from raw materials as it does now, therefore it is reasonable to assume that the chemicals that launched life on Earth were homochiral, just as their biogenically synthesized descendants are today. Even extraterrestrial amino acids arriving on Earth aboard meteorites are more or less homochiral. Do these compounds form in one isomer only or is some process destroying one isomer and not the other? The big difference between one chiral isomer and its opposite is its ability to absorb polarized light, leading scientists to suggest that it is polarized radiation from neutron stars that purifies substances in space into homochiral clouds.

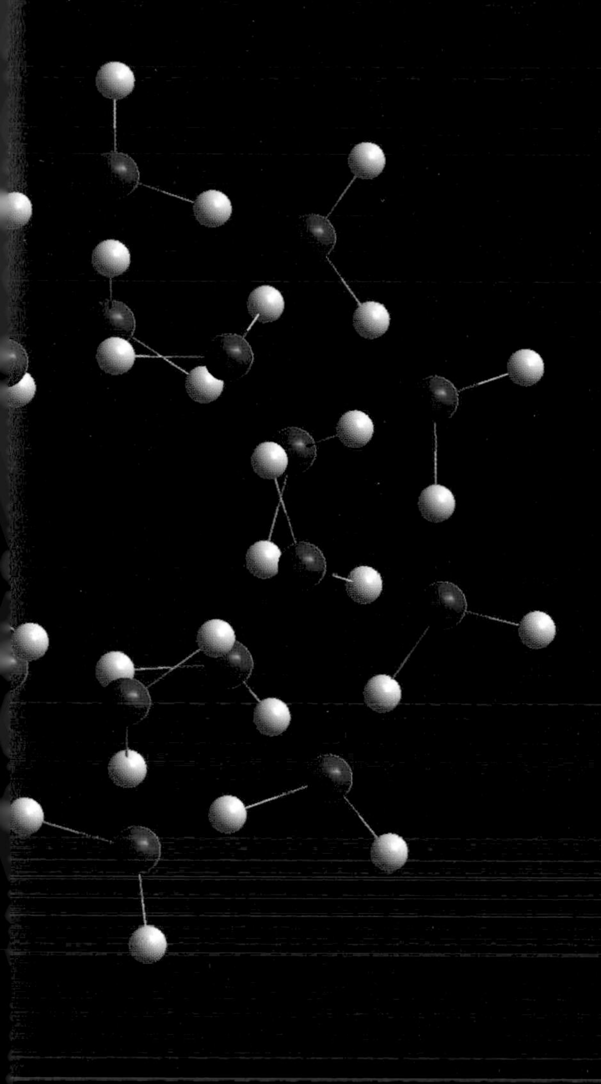

Can we build structures out of liquid water?

Earth is the only place in the Universe known to have liquid water on its surface. It is likely that other watery planets exist beyond our solar system but every world closer to home is either bone dry or is covered in ice. All the more unusual is that liquid water, perhaps the dominant compound in Earth—at least alongside silica—is still baffling chemists. One would have thought that this substance so central to our existence would have been thoroughly explained. However, the properties that make it so potent a chemical as a solvent, as a store of heat, in crystals, and in metabolism, also mean that its true nature is not fully understood. If it were, a whole new technology could result. Water molecules are polar, which means that the oxygen end is negatively charged, while the hydrogens are positively charged. This is the property that makes water very good at dissolving salts and other ionic compounds, which also have charged sections. Its polarity also means that water molecules link together with so-called *hydrogen bonds*. The simplest structure of water is called the dimer, with two molecules connected by a hydrogen bond, but in reality water molecules form clusters, constantly changing in complexity and structure. In effect, a sample of water is really a single cluster, a twisting and churning supermolecule. If chemists can learn to control the clusters then water-based structures could be made, including so-called water buckyballs, made from 28 molecules, and even an icosahedron of 280 molecules. It has been suggested they could solve climate change and even reveal the mystery of dark matter!

Does francium really exist?

The heaviest alkali metal, francium, is highly radioactive. It is formed by the decay of thorium or uranium, but its presence is only fleeting, with a half-life of just 20 minutes. Almost everything about this element is theoretical. It is thought to be the most reactive metal element of all, even more so than cesium, but its chemical properties are just inferred from those of others. It is so rare that the whole of Earth's rocks together are estimated to contain no more than 30g (just over one ounce) of francium at any one moment, with each of its atoms rapidly decaying only to be replaced for a few seconds by another coming into existence somewhere else. The largest sample of francium ever collected was 300,000 atoms, amassed in 2004. That was enough to form a ball of gas one millimeter across, that weighed 10 billionths of a gram and glowed with enough light to be seen—the one and only time francium has been. Scientists can make francium atoms, but is it really out there in nature?

The Great Chemists

Take a look at the lives behind the discoveries that shaped chemistry. More than a few of the protagonists were professional scientists working at universities ancient and modern or working with a commercial goal. Others came to the cutting edge as amateurs, men of means whose back-room tinkerings changed the world. Wherever the action happens, from the dappled groves above Athens to the atomic piles beneath Chicago, each one has a story to tell.

Aristotle

Born	384 BC
Birthplace	Greece
Died	322 BC
Importance	Most influential figure in early western science

The son of the king's doctor, Aristotle was born into Macedonian aristocracy. As befitting his status, he finished his education in Athens, as a pupil of Plato. Aristotle's legacy superseded that of his master and of any other Greek philosopher. It is easy to regard him as a hindrance to scientific enlightenment, since he got a great deal wrong. Nevertheless, Aristotle left us works on poetry, logic, metaphysics, language, and biology that gave intellectuals from as far afield as Turkmenistan and Ireland pause for thought for the best part of two millennia.

Arrhenius, Svante

Born	February 19 1859
Birthplace	Wik, Uppsala, Sweden
Died	October 2 1927
Importance	Ionic theory of reactions

It is said Arrhenius could read by the age of three—and he had taught himself. His autodidactic approach did not impress him to his teachers—his doctoral dissertation only just scraped a pass. It was this very work into the way compounds dissociated into charge ions that won him the Nobel Prize for Chemistry in 1903. Arrhenius was an influential figure in the Nobel Institute, which had been founded in 1900. He used his power to repeatedly block scientific rival and personal enemy Dmitri Mendeleev from receiving the top chemistry award.

Avogadro, Amadeo

Born	August 9 1776
Birthplace	Turin, Kingdom of Sardinia and Piedmont (Italy)
Died	July 9 1856
Importance	Avogadro's law of gases; Avogadro's number

As his name suggests (from Italian for lawyer), Avogadro was born into a long line of advocates. The young Amedeo was trained as an ecclesiastical lawyer but was soon devoting himself fully to science, dabbling in teaching, The article that was to add his name to the chemistry lexicon was published in 1811 but was before its time. When his home city of Turin fell under the rule of Sardinia after the fall of Napoleon, Avogadro became something of a political insurgent, resulting in his being ejected from his seat at the university. He was later reinstated but is still thought to have sponsored armed insurrection against the king.

Bacon, Roger

Born	c.1214
Birthplace	unknown
Died	1292
Importance	Promotion of experimental scientific method

In today's shorthand, Roger Bacon is often dubbed "one of the first scientists." In fact it is thought he did little actual hands-on research, instead collating and critiquing knowledge from other sources. Although his birth date is uncertain, Bacon is thought to have graduated from Oxford University at only 13. He later returned to his alma mater as a master and also held a similar post in Paris. In his forties he became a Franssciscan and is reported to have circumvented a publishing ban on friars by doing a deal with the then Pope, presenting his ideas in a theological context.

Becquerel, Henri

Born	December 15 1852
Birthplace	Paris, France
Died	August 25 1908
Importance	Pioneer in field of radioactivity

The Becquerels were a thoroughly scientific family. A Becquerel held the Applied Physics chair at France's Natural History Museum between 1838 and 1948—Henri was the third incumbent taking office in 1892, with his son Jean the fourth and final one. Becquerel also had a second career as the chief engineer for the French highways department. In 1903 he shared the Nobel Prize for Physics with the Curies for their work on radioactivity. Like many pioneers in this field, Becquerel did not live to be old. He died at the age of 55. The unit of radioactivity (Bq) is named for him.

Berzelius, Jöns Jacob

Born	August 20 1779
Birthplace	near Linköping, Sweden
Died	August 7 1848
Importance	Atomic weights; modern chemical symbols

The young Jöns Jacob could have chosen any career, but began by training as a doctor. Even so he could not resist introducing the latest science into his practice and spent several years giving sick people small electric shocks, to little advantage, if any. Fortunately for his patients, Berzelius's urge to research was redirected by a local mine owner who put him to work analyzing the minerals in the region, assessing them for their commercial worth. In 1808, Berzelius was given a position at the Karolinska Institute, one of Sweden's most distinguished schools, where he spent the rest of his professional career.

Black, Joseph

Born	April 16 1728
Birthplace	Bordeaux, France
Died	November 10 1799
Importance	Discovery of carbon dioxide

Joseph Black's parents were in the wine trade, and so as a child he would have been exposed to the natural and artificial chemical processes employed by vintners. He eventually settled on a career in medicine, and chemistry remained nothing more than a hobby. Even so he invented the analytical balance in the 1750s, a tipping scale that was accurate enough to weigh small quantities. Black was also a member of Scotland's literati of the day. He met regularly with Adam Smith and David Hume and was a close associate of James Watt.

Boyle, Robert

Born	January 25 1627
Birthplace	County Waterford, Ireland
Died	December 31 1691
Importance	Boyle's law on gases

Although there are several big names said to be the father of chemistry, Robert Boyle's 1661 book, *The Sceptical Chymist*, was an early attempt to put the study of the elements on a scientific footing by questioning the precepts of the alchemists. Aside from science, Boyle was also a man of god, investing his fortune in the East India Company to spread Christianity. A bequest in his will was made to fund lectures on the latest religious thinking. Despite a few hiatuses, the Boyle Lectures have been held yearly ever since.

Bunsen, Robert

Born	March 30 1811
Birthplace	Göttingen, Westphalia, Germany
Died	August 16 1899
Importance	Invention of the Bunsen burner

Robert Bunsen had a glittering scientific career. He is best remembered for his contribution of a new gas burner that now bears his name and the spectral analysis he used it for, which yielded the discovery of new elements in 1860. However, by then Bunsen was already a respected figure in the chemistry world, having made his name with the discovery of cacodyl, an explosive organic compound containing arsenic. This helped with the development of the valence theory, but Bunsen suffered for his success. He was blinded in one eye by one cacodyl explosion, and was almost poisoned by the arsenic.

Cannizzaro, Stanislao

Born	July 13 1826
Birthplace	Palermo, Italy
Died	May 10 1910
Importance	Development of Avogadro's theory on molecular weight

After a distinguished early career as a soldier and Sicilian politician, this Italian chemist still managed to have a chemical reaction named for him before he was 30. In the Cannizzaro reaction an aldehyde dissociates into its constituent carboxylic acid and alcohol, a major step forward in the processing of organic compounds such as those in petroleum. Cannizzaro is perhaps better remembered for his leading contribution to the Karlsruhe Congress in 1860, firing the imagination of Dmitri Mendeleev and giving rise to the Russian's periodic table.

Carothers, Wallace

Born	April 27 1896
Birthplace	Burlington, Iowa, U.S.A.
Died	April 29 1937
Importance	Invention of nylon

To please his father, Wallace Carothers trained first as an accountant but then devoted himself to an academic career in organic chemistry. In 1927, the DuPont chemical company approached Carothers to head up their new experimental research lab. His team produced neoprene, polyesters, and eventually nylon. However, despite these exciting successes and a doubling of his academic's salary, Carothers became depressed. On April 28, 1937, he killed himself by taking cyanide.

Cavendish, Henry

Born	October 10 1731
Birthplace	Nice, France
Died	February 24 1810
Importance	Identification of hydrogen gas

Born into a thoroughly aristocratic family, both of Henry Cavendish's grandfathers were Dukes. His lineage was a scientific one as well. His father was himself a member of the Royal Society of London, while a later cousin endowed Cambridge University with the Cavendish Laboratory, even today a leading research center. Cavendish was very shy and retiring. He had a private stairwell built on the back of his house so he could live a solitary existence, communicating with staff through notes. Nevertheless he was a regular at Royal Society dinners, but seldom spoke. As a result many of his discoveries were only revealed after he died.

Celsius, Anders

Born	November 27 1701
Birthplace	Uppsala, Sweden
Died	April 25 1744
Importance	100 degree scale thermometer

Best remembered for the temperature scale that bears his name, this Swedish astronomer was also a key figure in the French geodetic expedition to Lapland in 1736, which measured the shape of Earth using astronomical observations. While on his travels, Celsius recorded that the melting and boiling points of water were constant around the world. He used these as the upper and lower points in his temperature scale. Celsius originally designated boiling water as 0°C while the freezing temperature was 100°. Biologist Carl Linnaeus reversed these values after Celsius had died.

Curies, Marie and Pierre

Born	November 7 1867(Marie). May 15 1859(Pierre)
Birthplace (Marie)	Warsaw, Poland (then part of Russia)
Birthplace (Pierre)	Paris, France
Died	July 4 1934(Marie). April 19 1906(Pierre)
Importance	Pioneers in field of radioactivity

Marie Curie was born a Pole without a country. France represented an escape from German and Russian oppression in her homeland, where it was even illegal to speak Polish. Marie worked to fund her sister's education in Paris and then she followed, taking two degrees at the Sorbonne. She met Pierre a few years later. He had already made a discovery: magnets lost their force above a critical temperature. Both suffered from bad radiation burns, with Marie dying of leukemia probably induced by radioactivity.

Dalton, John

Born	September 5 or 6 1766
Birthplace	Eaglesfield, Cumberland, England
Died	July 27 1844
Importance	Modern atomic theory

John Dalton was a Quaker, and although he himself had been a teacher since the age of 15, he was barred from entering British universities, which did not allow members of dissenting churches. Dalton was mostly self-taught and received informal schooling from John Gough, a Manchester polymath. Dalton was color blind and made one of the first formal descriptions of the condition in 1794. He lived frugally in Manchester, England, even after election to the Royal Society of London. The unit of atomic mass is named dalton (Da) in his honor. One dalton is one twelfth of the mass of an atom of carbon 12.

Davy, Humphry

Born	December 17 1778
Birthplace	Penzance, Cornwall, England
Died	May 29 1829
Importance	Pioneer of electrolysis

Davy, the son of a poor Cornish woodcarver, owed his start in life to a wealthy Penzance doctor who had been his mother's guardian. Davy continued to take every opportunity that came his way and his reputation spread among both the scientific and socialite communities. As a new member of the Royal Society in the early 1800s, he attended a performance given by Luigi Galvani's nephew, showing the effects of electricity on recently executed convicts. Davy is reported to have told his friend Mary Shelley about this event, and this supposedly was one of the inspirations for her 1818 novel *Frankenstein*.

Democritus

Born	c.460 BC
Birthplace	A Greek colony in modern-day Turkey
Died	c.370 BC
Importance	One of the "founding fathers of science;" atomic theory of the universe

Democritus certainly qualifies as a "founding father of science" along with the likes of Thales and Aristotle. He was born in a Greek colony in what is now west Turkey, and is reported to have traveled farther and wider than his contemporaries. His work was inspired by lessons with Egyptian mathematicians, the magi of Persia, and the astronomers of Babylon. Nevertheless, he lived a modest life, refusing to court fame. One story attests that Democritus blinded himself into order to increase the focus of his thoughts. He appears to have had a humorous outlook, often joking his way through critiques of other philosophies.

Faraday, Michael

Born	September 22 1791
Birthplace	Newington, Surrey, England
Died	August 25 1867
Importance	Discovery of electromagnetic induction

Born on the outskirts of London, Michael Faraday was apprenticed as a bookbinder. However, a visit to the Royal Institution to hear lectures by Humphry Davy and others set his ambitions on a different track. His notes on what he had learned from the lectures so impressed Davy that Faraday eventually became his assistant. Faraday's own research achievements resulted in a conflict with his powerful mentor, which is thought to be one of the reasons that sent Faraday into a deep depression in middle age. By later life he was held in high esteem by the British people, but did little research.

Fermi, Enrico

Born	September 29, 1901
Birthplace	Rome, Italy
Died	November 2 1954
Importance	Construction of first nuclear reactor (atomic pile)

Enrico Fermi's scientific prowess was born out of tragedy. His brother died young, and the teenager Enrico battled grief with study. At the age of 24, Fermi became Italy's first professor of atomic physics. Within a decade he had opened the door to unlimited nuclear power. He went to Sweden to collect a Nobel prize in 1938, but did not go back to Rome. As a Jew in a Europe gripped by fascism, Fermi thought it better to continue his work into nuclear fission in the U.S. Like many of his colleagues, Fermi died of cancer. In the early days of atomic physics, the dangers of radioactivity were not well known.

Gay-Lussac, Joseph-Louis

Born	December 6 1778
Birthplace	Saint-Léonard-de-Noblat, France
Died	May 9 1850
Importance	Gay-Lussac's law

Perhaps fortunate to be just a child during the French Revolution, Gay-Lussac avoided much of the turmoil that plagued scientific figures of the period—although his father was arrested during the Reign of Terror. While only a demonstrator at a Paris college, he discovered the gas law named for him. In 1804 he ascended 7,000 metres into the air in a hot-air balloon to collect samples of air at different altitudes. He often collaborated with Alexander von Humboldt, famous for contributions to geography and biology. Gay-Lussac also invented the terms pipette and burette.

Hooke, Robert

Born	July 18 1635
Birthplace	Isle of Wight, England
Died	March 3 1703
Importance	Hooke's law of elasticity

Robert Hooke is a much overlooked scientist. He was one of Robert Boyle's assistants, doing important work on the air pumps crucial to his employer's research. Hooke was also a key figure in the foundation of the Royal Society of London in the 1660s and is best remembered for Hooke's law of elasticity which is used to describe everything from bungee jumps to the vibration of atoms. Hooke was also one of the first scientists to turn a microscope on biological specimens. He reported seeing small enclosures within the tissue of plants, which he likened to the rooms of monks, naming them "cells."

Haber, Fritz

Born	December 9 1868
Birthplace	Breslau, Prussia (now Wroclaw, Poland)
Died	January 29 1934
Importance	Nitrogen fixation for fertilizers; chemical weapons

Fritz Haber was born a Jew in a German-speaking village in what is now Poland, but later converted to Lutheranism. His personal life was forever tarnished by his work in chemistry: not his contribution to nitrogen fixation, but his work on chemical weapons. In April 1915 he oversaw the release of chlorine gas during the Battle of Ypres. His first wife Clara shot herself a month later. It of course must be remembered that Haber was a military officer in a war for national survival against enemies that were deploying their own chemical weapons.

Hypatia of Alexandria

Born	c.355
Birthplace	Alexandria, Egypt
Died	March 415
Importance	Invention of the hydrometer

Hypatia was a woman who made her mark at a time when women seldom made it into a classroom let alone into history books. Hypatia's father was the last head of the great Library of Alexandria, and so she had the best education from the start. She is credited with the invention of the hydrometer, a glass and mercury weight that floated in liquids. The height at which it floated indicated the liquid's density. Hypatia was one of the last classical scholars and was murdered by Christians who objected to her teaching Plato's heretical philosophy.

Klaproth, Martin

Born	December 1, 1743
Birthplace	Wernigerode, Brandenburg
Died	January 1 1817
Importance	Discovery of uranium and other elements

This German chemist was one of the most prolific discoverers of new elements in history. While others investigated air, he focused on the analysis of minerals, discovering uranium, titanium (with others), and zirconium. He also found evidence for tellurium, strontium, cerium, and chromium. Klaproth spent the first half of his career as a pharmacist in various German cities; chemistry was just a hobby. By the 1780s, however, after starting a business in Berlin, he was appointed to a position in the city's military college, later becoming the professor of chemistry when it converted to a full university.

Krebs, Hans

Born	August 25 1900
Birthplace	Hildesheim, Germany
Died	November 22 1981
Importance	Discovery of the citric acid cycle

Hans Krebs was a German-born physician. In 1932 he succeeded in figuring out the urea cycle, the means by which mammals remove waste proteins in their urine. In the same year he joined the German army, but as a Jew he was forced out the following year and barred from private medical practice. He emigrated to England, taking a research post at the university of Cambridge. He later moved to Sheffield, becoming the biochemistry professor there in 1945. Krebs was awarded the Nobel Prize for Medicine in 1953 and was knighted by the British monarch in 1958, becoming Sir Hans Krebs.

Laplace, Pierre-Simon

Born	March 23 1749
Birthplace	Normandy, France
Died	March 5 1827
Importance	Coinventor of calorimeter

As well as helping fellow aristocrat Lavoisier in his research into the nature of heat, Pierre-Simon Laplace was also a prolific mathematician and astronomer. He is best remembered for his work on probability theory. While Lavoisier fell foul of the revolutionary leaders of France,

Laplace steered a different course, and became one of Napoleon Bonaparte's (the eventual French emperor) scientific advisers. Laplace was a stout defender of reason. When asked by Napoleon why God was not mentioned in his work, Laplace replied, "I had no need of that hypothesis."

Lavoisier, Antoine

Born	August 26 1743
Birthplace	Paris, France
Died	May 8 1794
Importance	Discovery of the composition of water

Antoine Lavoisier was a momentous figure in chemistry, naming oxygen and hydrogen gases and showing how they formed water, until then thought an element. Lavoisier's life coincided with the French Revolution. He had inherited a fortune from his mother and used it to buy a stake in a tax-farming company, drawing a substantial income for collecting taxes on behalf of the hated king, Louis XVI. Despite suggesting reforms to help the peasant classes prior to the revolution, and working for the new state afterward, Lavoisier's tax job proved his downfall, and he was sent to the guillotine.

Lawrence, Ernest O

Born	August 8 1901
Birthplace	Canton, South Dakota, U.S.A.
Died	August 27 1958
Importance	Invention of the particle accelerator

Ernest Lawrence was the inventor of the cyclotron, the first particle accelerator used to smash atoms together at immense speeds. His work was instrumental in the formation of the first technetium samples, and similar experiments revealed many more artificial elements. Lawrence received a Nobel prize in 1939 and his expertise was used during the Manhattan Project to develop methods of separating the fissile isotopes from natural uranium. This created a sample for use in the first atomic weapons. Element 103 was named lawrencium in his honor in 1963.

Mary the Jewess

Born	Between 1st and 3rd centuries AD
Birthplace	?
Died	Between 1st and 3rd centuries AD
Importance	Invention of the bain-marie

Also known as Maria the Prophetess, Mary of Alexandria and, chiefly in Arabic sources, Daughter of Plato, Mary is a mysterious figure. It is not even certain that she was Jewish rather a Coptic Christian. It is unclear exactly when she lived, but it is safe to rule out that she was Moses' sister, which is one of many claims. Mary's lasting contribution was in apparatus: her water bath, or *bain marie*, was a means of heating substances gently. Her *tribikos* and *kerotakis* were distillation devices.

Leucippus

Born	flourished 5th century BC
Birthplace	Probably at Miletus, Asia Minor
Died	5th century BC
Importance	Development of the theory of atomism

While Democritus is best known for expanding and explaining the concept of atoms, it was probably his teacher Leucippus who came up with the idea first. Little is known about this Greek philosopher. He is thought to have hailed from Miletus, a city with a strong heritage of science-minded philosophy. Aristotle credits the theory of atomism to him, which was taken on by Democritus around 430 BC. Leucippus reasoned that the Universe was constructed of two elements, the full and the empty—solid and vacuum. It was the interaction of the two that created natural phenomena.

Mendeleev, Dmitri

Born	January 27 (February 8, New Style) 1834
Birthplace	Tobolsk, Siberia, Russia
Died	January 20 (February 2) 1907
Importance	Inventor of the periodic table

Dmitri Mendeleev was born in a Siberian village, the youngest of a large family—most estimates put the number of siblings above 12. In the 1850s, the family moved to St Petersburg. There Dmitri had access to a better education and won a place in Heidelberg as a student of Robert Bunsen. In the mid-1860s he became established at St Petersburg's university where he remained for the rest of his career. A tempestuous love life, including allegations of bigamy, hindered his advancement in the Russian Science Academy but he was much lauded elsewhere.

Nobel, Alfred

Born	October 21 1833
Birthplace	Stockholm, Sweden
Died	December 10 1896
Importance	Invention of dynamite; Nobel Prizes

Alfred Nobel's name is synonymous with the greatest achievements in science, medicine, economics, and politics through the prizes awarded by an institute bequeathed in his will. He did this to leave a better legacy than the invention of dynamite and gelignite, for which he was responsible. These inventions, combined with his acquisition of the steel and weapons firm Bofors, were the source of his great wealth. Nobel was a chemist by training. His success with dynamite was by finding a way to stabilize the explosive component with fine-grained earth.

Pasteur, Louis

Born	December 27 1822
Birthplace	Dole, France
Died	September 28 1895
Importance	Discovery of chirality

Despite making big discoveries in molecular isomerism, Louis Pasteur is considered one of the founders of microbiology. He is most often remembered for his work on food safety, developing the pasteurization method of removing germs from liquid foodstuffs, such as milk and wine. He discovered that "germs"—bacteria and other microorganisms—were the cause of disease and decay through a series of famous experiments involving broth in swan-necked flasks, some sealed, some exposed to the air and the germs it contained.

Ørsted, Hans Christian

Born	August 14 1777
Birthplace	Rudkøbing, Denmark
Died	March 9 1851
Importance	Discovery of electromagnetism

Hans Christian Ørsted was largely home-schooled and did so well at the University of Copenhagen that he was awarded a travel scholarship in 1801, giving him the opportunity to tour Europe for three years to learn from the greatest thinkers of the time. Following this journey, Ørsted put other interests aside and focused on the study of physics and chemistry. His breakthrough in linking electricity and magnetic fields in 1820 had far-reaching effects in both sciences. The unit of magnetic induction is named the oerstead (Oe) in his honor.

Pauling, Linus

Born	February 28 1901
Birthplace	Portland, Oregon, U.S.A.
Died	August 19 1994
Importance	Developed theory of covalent bonding

Linus Pauling excelled in high school and by the age of 16 had enough credits to enter Oregon State University (then an agricultural college). Although he had not passed the correct history course to graduate high school, he left anyway. He finally got his school diploma in the 1960s after winning two Nobel prizes. Pauling declined involvement in the Manhattan Project for family reasons during World War II, but became involved in other military research. Along with Albert Einstein, he was soon campaigning against nuclear weapons after the war's end.

Plato

Born	428/427 BC
Birthplace	Athens, Greece
Died	348/347 BC
Importance	Philosophy; theory of "ideas" or "forms"

Plato was taught by Socrates, and most of what we know of this great philosopher is from his pupil's accounts. Plato later founded the Akademia, a school of Athenian philosophy that dominated western thought. The name, now so associated with schooling, probably derives from the previous owner of the land. One of its famous pupils was Aristotle. Plato's given name was Aristocles, but given his link to Aristotle, it is perhaps fortunate for history that he was given the nickname Plato, meaning "broad," by his wrestling teacher.

Seaborg, Glenn

Born	April 19 1912
Birthplace	Ishpeming, Michigan, U.S.A.
Died	February 25 1999
Importance	Discovery of transuranium elements

Glenn Seaborg holds a unique position among scientists, one that is likely to remain his alone. After being involved in the discovery of ten artificial elements, Seaborg was given the ultimate honor of having element 106 named for him. seaborgium became its official name in 1997, the only time an element has had a still living namesake. Seaborg was a crucial researcher in the 1940s Manhattan Project that created atomic weapons, but was also a negotiator in many of the Cold War non-proliferation treaties that followed in the decades after.

Rutherford, Ernest

Born	August 30 1871
Birthplace	Spring Grove, New Zealand
Died	October 19 1937
Importance	Discovery of the atomic nucleus

Born on a humble farm on New Zealand's North Island but later elevated to Baron Rutherford of Nelson, Ernest Rutherford's name appears throughout the early history of atomic physics. Chadwick, Geiger, Bohr, Hahn, and Soddy all worked under him at some point, often directed by his theories toward their own personal discoveries. He was buried in London's Westminster Abbey near Isaac Newton and other great British scientists. In 1997, element 104 was named rutherfordium (Rf) in his honor.

Thomson, J.J.

Born	December 18 1856
Birthplace	Manchester, England
Died	August 30 1940
Importance	Discovery of electrons

The founding father of particle physics, Joseph John Thomson revealed that atoms were not indivisible solids but constructed of yet smaller particles. Rutherford was counted among his students. An excellent school pupil, his parents planned for Thomson to be apprenticed as a steam engine mechanic, but he was admitted to Trinity College, Cambridge, to read mathematics and physics at just 17 years old. He never really left, becoming the professor of physics there in 1884. Thomson's son George also won a Nobel prize for researching the wave-particle duality of electrons, the very particles discovered by J.J.

BIBLIOGRAPHY AND OTHER RESOURCES

Books

Atkins, P.W. *Galileo's Finger: The Ten Great Ideas of Science*. Oxford: Oxford University Press, 2004.

Cobb, Cathy and Harold Goldwhite. *Creations of Fire: Chemistry's Lively History from Alchemy to the Atomic Age*. New York: Plenum Publishing, 1995.

Emsley, John. *Nature's Building Blocks: An A-Z Guide to the Elements*. Oxford: Oxford University Press, 2011.

Gray, Theodore. *The Elements: A Visual Exploration of Every Known Atom in the Universe*. New York: Black Dog & Leventhal Publishers, 2009.

Kean, Sam. *The Disappearing Spoon: And other true tales from the Periodic Table*. London: Black Swan, 2010.

MacArdle, Meredith (ed.). *Scientists: Extraordinary People who Changed the World*. London: Basement Press, 2008.

Scerri, Eric R. *The Periodic Table: A Very Short Introduction*. Oxford: Oxford University Press, 2011.

Stwertka, Albert. *A guide to the elements*. New York: Oxford University Press, USA, 2012.

Suplee, Curt. *Milestones of Science*. Washington, D.C.: National Geographic Society, 2000.

Chemical Societies

American Chemistry Society www.acs.org

Chemical Institute of Canada www.cheminst.ca

Chemical Society of Japan www.csj.jp

Chinese Chemical Society www.ccs.ac.cn

European Association for Chemical and Molecular Sciences www.euchems.org

Gesellschaft Deutscher Chemiker (German Chemical Society) www.gdch.de

Indian Chemical Society www.indianchemsoc.org

Mendeleev Russian Chemical Society, Moscow, Russia

Royal Australian Chemical Institute www.raci.org.au

Royal Society of Chemistry www.rsc.org

Societá Chimica Italiana (Italian Chemical Society) www.soc.chim.it

Société Chimique de France (Chemical Society of France) www.societechimiquedefrance.fr

Swedish Chemical Society www.chemsoc.se

Museums

Calalyst, Cheshire, England, U.K. www.catalyst.org.uk

Chemical Heritage Foundation, Philadelphia, Pennsylvania, U.S.A. www.chemheritage.org

China Science and Technology Museum, Beijing, China www.cstm.org.cn

Cité des Sciences et de l'Industrie, Paris, France www.cite-sciences.fr

Copernicus Science Centre, Warsaw, Poland www.kopernik.org.pl/en/

Deutsches Museum, Munich, German www.deutsches-museum.de

Exploratorium, San Francisco, U.S.A. www.exploratorium.edu

Museo della Scienza e dalla Tecnologia "Leonardo da Vinci," Milan, Italy www.museoscienza.org

Museo Galileo, Florence, Italy www.imss.fi.it

Museum of Science, Boston, U.S.A. www.mos.org

Museum of Science and Industry, Chicago, U.S.A. www.msichicago.org

National Museum of Nature and Science, Tokyo, Japan www.kahaku.go.jp/english/

Norwegian Museum of Science and Technology, Oslo, Norway www.tekniskmuseum.no

Observatory Museum, Stockholm, Sweden www.observatoriet.kva.se/engelska

Ontario Science Centre, Toronto, Canada www.ontariosciencecentre.ca

Powerhouse Museum, Sydney, Australia www.powerhousemuseum.com

Saint Louis Science Center, St. Louis, U.S.A. www.slsc.org

Science Center NEMO, Amsterdam, Netherlands www.e-nemo.nl/en

Science Museum, London, U.K. www.sciencemuseum.org.uk

Shanghai Science and Technology Museum, Shanghai, China www.sstm.org.cn

Smithsonian Institution, Washington D.C., U.S.A. www.si.edu

Archives or Preserved Equipment

Berzelius Exhibition, The Observatory Museum, Stockholm, Sweden www.observatoriet.kva.se

Boyle Archive/Hooke Papers/Davy Papers, Royal Society, London, U.K. www.royalsociety.org

Bunsen Archive, Royal Society of Chemistry, London, U.K. www.rsc.org

Carothers and du Pont Papers, Hagley Museum and Library, Wilmington, Delaware, U.S.A. www.hagley.org

Gay-Lussac Papers, École Polytechnique, Paris, France www.polytechnique.edu

Humphry Papers/Faraday Notebooks, Royal Institution of Great Britain, London, U.K. www.rigb.org

Krebs Papers, University of Sheffield Library, Sheffield, England, U.K. www.shef.ac.uk/library

Lavoisier Laboratory, Museé des Arts et Métiers, Paris, France www.arts-et-metiers.net

Liebig Museum, Giessen, Germany www.liebig-museum.de

Maria Skłodowska-Curie Museum, Warsaw, Poland www.muzeum-msc.pl

Mendeleev Museum and Archives, St.Petersburg, Russia www.russianmuseums.info/M124

Nobel Archive, National Archives, Stockholm, Sweden www.riksarkivet.se

Pasteur Museum, Paris, France www.pasteur.fr

Pasteur Museum in Dole, France www.musee-pasteur.com

Pauling Papers, Oregon State University, U.S.A. www.pauling.library.oregonstate.edu

Rutherford Museum, McGill University, Montreal, Canada www.physics.mcgill.ca/museum/rutherford_museum.htm

Websites

Dynamic periodic table: www.ptable.com

Khan Academy: www.khanacademy.org/#chemistry

Nobel Foundation: www.nobelprize.org

INDEX

Cataloging-in-Publication Data has been applied for and may be obtained from the Library of Congress.

ISBN 978-0-9853230-3-5

Series Concept and Direction: Jeanette Limondjian
Design: Bradbury and Williams
Editor: Meredith MacArdle
Picture Research: Jennifer Veall
Cover Design: Jokooldesign

Publisher's Note: While every effort has been made to insure that the information herein is complete and accurate, the publishers and authors make no representations or warranties either expressed or implied of any kind with respect to this book to the reader. Neither the authors nor the publisher shall be liable or responsible for any damage, loss or expense of any kind arising out of information contained in this book. The thoughts or opinions expressed in this book represent the personal views of the authors and not necessarily those of the publisher. Further, the publisher takes no responsibility for third party websites or their content.

SHELTER HARBOR PRESS
603 West 115th Street Suite 163
New York, New York 10025

Printed and bound in China by Imago.

10 9 8 7 6 5 4 3 2 1

PICTURE CREDITS

BOOK

Alamy/The Print Collector 48; INTERFOTO 50 top; Keystone Pictures USA 103 bottom left; The Natural History Museum 135 top right. **Bradbury and Williams** 93 right, 94 bottom right, 95 bottom left, 112 bottom left, 119 bottom; 120, 121, 123, 124-5, 127 top left, 129 bottom right. **Chemical Heritage Foundation**/Osnovy khimii, 1869-1871, Dmitri Ivanovich Mendeleev, Roy G. Neville Historical Chemical Library, Photograph by Douglas A. Lockard 7 top; Liber de arte Distillandi, 1512, Hieronymus Brunschwig, Othmer Library of Chemical History, Photograph by Gregory Tobias 30 top; Portrait of Joseph Priestley, Attributed to Ozias Humphrey, Gift of Chemists' Club, Photograph by Will Brown 44; The Shannon Portrait of the Hon. Robert Boyle, F.R.S. (1627-1691), 1689, Johann Kerseboom, purchased with funds from Eugene Garfield and the Phoebe W. Haas Charitable Trust, Photograph by Will Brown 132 top left. **Churchill Archives Centre** 99. **Corbis**/The Gallery Collection 36; Hemis 51; Raymond Reuter/Sygma 64 right; Roger Ressmeyer 109. **Dr. Keith Johnson, MIT & Watercluster Sciences Inc.** 128-129. **Edgar Fahs Smith Image Collection, Rare Book and Manuscript Library, University of Pennsylvania** 30 bottom, 33, 43 top, 50 bottom, 62, 68 bottom, 72, 130 bottom right, 132 top right, 132 bottom left, 136 top left, 137 bottom right, 138 top right. **Getty Images**/SSPL via Getty Images 15 bottom, 17, 68 top, 89 bottom; AFP 115. © **Heinrich Pnoik** 80-81 bottom. **Institut International de Physique Solvay** 94 top. **Library of Congress** 55 bottom. **NASA**/JPL-Caltech, UCLA 108 top; 128 bottom. **Science Photo Library**/RIA Novosti endpapers, 2-3; Pasieka 2-3; Sheila Terry 4 background; Gregory Tobias, Chemical Heritage Foundation 4 bottom; 5 background; NYPL, Science Source 6 left; Sheila Terry 6 right; Martin Land 11; Scientifica, Visuals Unlimited 12 top; Pasquale Sorrentino 14 bottom; Royal Astronomical Society 18; Albert Copley, Visuals Unlimited 19 top; Middle Temple Library 19 bottom; National Library of Medicine 20 top; Cordelia Molloy 20 bottom; 21 top; CCI Archives 21 bottom; 22; Emilio Segre Visual Archives/American Institute of Physics 23 top; Martyn F. Chillmaid 27; Jean-Loup Charmet 28 top; Middle Temple Library 28 bottom; New York Public Library Picture Collection 29; 31; 32 left; Gregory Tobias, Chemical Heritage Foundation 34 left; Royal Astronomical Society 34-35; New York Public Library Picture Collection 35, 38-39, 40 top; Sheila Terry 41; US Navy 43 bottom; 45 top; 45 bottom; 46 top; 47 top; 47 bottom; Sheila Terry 49 top; Gregory Tobias, Chemical Heritage Foundation 49 bottom; Science Source 52; Sheila Terry 53 right; Andrew Lambert Photography 57; 59 top; New York Public Library Picture Collection 59 bottom; 61 left; Sheila Terry 61 right; Royal Institution of Great Britain 63 top; 63 bottom; Jacopin 65; Friedrich Saurer 66 top left; Andrew Lambert Photography 66 top right; Charles D. Winters 66 bottom; Detlev van Ravenswaay 70 top; NYPL, Science Source 70 bottom; 73; National Library of Congress 75 top; 76 top; Charles D. Winters 76 bottom; 77 top; 77 bottom; Emilio Segre Visual Archives/American Institute of Physics 78; 79 top; Jose Antonio Peñas 79 bottom; 81; Physics Today Collection/American Institute of Physics 83; Prof. Peter Fowler 84 top; Prof. K. Seddon & Dr. T. Evans. Queen's University Belfast 84 bottom; Science Source 85; 86; Prof. Peter Fowler 88-89; Rick Miller/Oxford Centre for Molecular Sciences 90 left; National Physical Laboratory © Crown Copyright 90 right; Patrick Landmann 91; Lawrence Berkeley Laboratory 92 bottom; Jean-Claude Revy, ISM 94 bottom left; Science Source 96; Kenneth Eward, Biografx 97; Thomas Hollyman 98 bottom; Eye of Science 100; David Parker 101; Prof. K. Seddon & Dr. T. Evans. Queen's University Belfast 103 top; Gary Sheahan/US National Archives and Records Administration 104 bottom; US National Archives and Records Administration 104-105; Lawrence Berkeley National Laboratory 106; U.S. Dept. of Energy 107 bottom; Fred McConnaughey 108 bottom; A. Barrington Brown 110; Laguna Design 112 top; Pasieka 113 left; Physics Dept., Imperial College 114 top; Andy Crump 114 bottom; Victor Habbick Visions 115 bottom; Lawrence Livermore National Laboratory 116 top; Roger Harris 116-117; Detlev van Ravenswaay 117 top; Adam Hart-Davis 117 bottom; Charles D. Winters 118 left; 118 right; Claus Lunau 119 top; Pasieka 121 bottom; 122 bottom; Ted Kinsman 126 top; Laguna Design 126 bottom; Victor Habbick Visions 127; 130 top left; Middle Temple Library 130 bottom left; 131 top right; National Library of Congress 131 bottom left; 132 bottom right, 133 top left; American Institute of Physics 133 top right; New York Public Library Picture Collection 133 bottom left; Paul D. Stewart 134 top left; National Library of Medicine 134 bottom left; American Institute of Physics 134 bottom right; Royal Institution of Great Britain 135 top left; 135 bottom left, 135 bottom right, 136 top right; US National Library of Medicine 136 bottom left; Lawrence Berkeley Laboratory 137 top left; Middle Temple Library 137 top right; Science Source 137 bottom left; 138 top left; Thomas Hollyman 138 bottom right; Middle Temple Library 139 top left; Lawrence Berkeley National Laboratory 139 top right; Emilio Segre Visual Archives/American Institute of Physics 139 bottom left; 139 bottom right. **Taken from *Traite Elementaire de Physique*, Augustin Privat Deschanel (1869)** 67. **Taken from D-Kuru/www.commons.wikimedia.org/wiki/File:Crookes_tube_two_views.jpg** 74 bottom. **Thinkstock**/Dorling Kindersley RF 5 top; Hemera 10-11; Comstock 12 bottom; iStockphoto 13; Photos.com 14-15; Hemera 16 top; iStockphoto 16 bottom; Photos.com 24, 25 bottom, 26; iStockphoto 27 bottom; Photos.com 32 right; Digital Vision 37; Hemera 39; Photos.com 40 bottom, 42; iStockphoto 46 bottom; Photos.com 53 left, 55 top, 60, 69; iStockphoto 71, 74 top, 75 bottom; Photos.com 82 top, 82 bottom; Hemera 87 top; Dorling Kindersley RF 92 top; Hemera 104 top, 105 top; Dorling Kindersely RF 105 bottom; Digital Vision 107 top; Hemera 111; iStockphoto 113 right, 122 top; Photos.com 131 top left, 131 bottom right, 133 bottom right, 134 top right; iStockphoto 136 bottom right; Photos.com 138 bottom left. **Colin Woodman** 65 bottom, 78 top, 80 top, 85 top, 95 bottom right.

TIMELINES

Alamy/AF archive; D. Hurst, epa european pressphoto agency b.v.; France Roberts; Images & Stories; INTERFOTO; Mary Evans Picture Library; The Print Collector; Vintage Images; ZUMA Wire Service. **Science Photo Library**/A. Barrington Brown; Adam Jones; Carolyn Brown; CCI Archives; Charles D. Winters; Chris Gallagher; Custom Medical Stock Photo; David Hardy; David R. Frazier; Detlev van Ravenswaay; Eye of Science; Gary Cook/Visuals Unlimited, Inc.; George Bernard; Hank Morgan; Hubertus Kanus; James King-Holmes; John Heseltine; Library of Congress; Maria Platt-Evans; Michael Gilbert; Miriam And Ira D. Wallach Division of Art, Prints and Photographs/New York Public Library; NASA; Natural History Museum, London; NYPL; Pasieka; Patrick Landmann; Photo Researchers; Power and Syred; RIA Novosti; Royal Astronomical Society; Science Source; Sheila Terry; Simon Fraser; Take 27 Ltd.; Tek Image; University of Chicago/American Institute of Physics; USA National Library of Medicine; US Army; U.S. Dept. of Energy. **Taken from Xuan Che/www.commons.wikimedia.org /wiki/File: The_mask_of _ agamemnon .jpg**. **Thinkstock**/Hemera; iStockphoto; Paul Katz; Photos.com.

Publisher's note: Every effort has been made to trace copyright holders and seek permission to use illustrative material. The publishers wish to apologize for any inadvertent errors or omissions and would be glad to rectify these in future editions.